Horizons in World Physics
Volume 239

THEORETICAL PHYSICS 2002, PART 1

Horizons in World Physics

Volume 239	**Theoretical Physics 2002, Part 1** Henk F. Arnoldus and Thomas F. George
Volume 238	**Theoretical Physics 2001** Henk F. Arnoldus and Thomas F. George
Volume 237	**Generalized functions in mathematical physics: Main ideas and concepts** A.S. Demidov
Volume 236	**Edge Excitations of Low-Dimensional Charged Systems** Oleg Kirichek
Volume 235	**On the Structure of the Physical Vacuum and a New Interaction in Nature (Theory, Experiment, Applications)** Yu. A. Baurov
Volume 234	**Electron Scattering on Complex Atoms (Ions)** V. Lengyel, O. Zatsarinny and E. Remeta
Volume 233	**An Introduction to Hot Laser Plasma Physics** A.A. Andreev, A.A. Mak and N.A. Solovyev
Volume 232	**Electron-Atom Scattering: An Introduction** Maurizio Dapor
Volume 231	**Theory of Structure Transformations in Non-equilibrium Condensed Matter** Alexander L. Olemskoi
Volume 230	**Topics in Cosmic-Ray Astrophysics** M. A. DuVernois (Ed.)
Volume 229	**Interactions of Ions in Condensed Matter** A. Galdikas and L. Pranevièius
Volume 228	**Optical Vortices** Edited by M. Vasnetsov and K. Staliunas
Volume 227	**Frontiers in Field Theory, Quantum Gravity and Strings** R.K. Kaul, J. Maharana, S. Mukhi and S. K. Rama
Volume 226	**Quantum Groups, Noncommutative Geometry and Fundamental Physical Interactions** D. Kastler, M. Russo and T. Schücker
Volume 225	**Aspects of Gravitational Interactions** S.K. Srivastava and K.P. Sinha
Volume 224	**Dynamics of Transition Metals and Alloys** S. Prakash
Volume 223	**Fields and Transients in Superhigh Pulse Current Devices** G.A. Shneerson
Volume 222	**Gaussian Beams and Optical Resonators** A. N. Oraevsky
Volume 221	**Laser Cathode-Ray Tubes** Edited by Yu. M. Popov
Volume 220	**Quantum Electrodynamics with Unstable Vacuum** Edited by V.L. Ginzburg
Volume 219	**The Muon Method in Science** V.P. Smilga and Yu.M. Belousov
Volume 218	**Radiative and Collisional Characteristics of Ions in Hot Plasmas** Edited by I.I. Sobel'man
Volume 217	**Nonlinear and Quantum Optical Phenomena in Nonequilibrium Media** Edited by V.A. Shcheglov
Volume 216	**Semiconductors and Insulators: Optical and Spectroscopic ...** Edited by Yu.I. Koptev
Volume 215	**Relativistic Theory of Gravity** A.A. Logunov
Volume 214	**Research in Quantum Field Theory** Edited by V.I. Man'ko
Volume 213	**Nonlinear Theory of Strong Electromagnetic Wave-plasma...** Edited by O.N. Krokhin

Volume 212 **Dynamics of Elementary Atomic-Molecular...**
Edited by O. S. Scheglov

Volume 211 **Optics and Lasers**
Edited by G.G. Petrash

Volume 210 **Theory and Technology of High Temperature Superconductors**
Edited by Yu.I. Koptev

Volume 209 **Theory of Interaction of Multilevel Systems...**
Edited by V.I. Man'ko

Volume 208 **Physical Properties of High Temperature Superconductors**
Edited by Yu.I. Koptev

Volume 207 **Generalized Mayer Series in Classical Statistical Mechanics**
Igor J. Ivanchik

Volume 206 **Mathematical Physics, Applied Mathematics and Informatics**
Edited by Yu.I. Koptev

Volume 205 **Squeezed and Correlated States of Quantum Systems**
Edited by M.A. Markov

Volume 204 **Sakharov Memorial Lectures in Physics**
Edited by L.V. Keldysh and V.Ya. Fainberg

Volume 203 **Fiber Optics: Research and Development**
Edited by Ye.M. Dianov

Volume 202 **Gas Dynamics**
Edited by Yu.I. Koptev

Volume 201 **Generation of Nonlinear Waves and Quasistationary ...**
Edited by L.M. Kovrizhnykh

Volume 200 **Laser Thermonuclear Targets and Superdurable Microballoons**
Edited by A.I. Isakov

Volume 199 **Spatiotemporal Characteristics of Laser Emission**
M.V. Pyatakhin and A.F. Suchkov

Volume 198 **Theory of Nonstationary Quantum Oscillators**
Edited by M.A. Markov

Volume 197 **Reliability Problems of Semiconductor Lasers**
P.G. Eliseev

Volume 196 **Pulsars**
Edited by A.D. Kuzmin

Volume 195 **Physical Characteristics and Critical Temperature of High Temperature Superconductors**
Edited by M.M. Sushchinskiy

Volume 194 **Biophysical Approach to Complex Biological Phenomena**
Edited by E. Volkov

Volume 193 **Research on Chemical Lasers**
Edited by A.N. Orayevskiy

Volume 192 **Atomic and Ionic Spectra and Elementary Processes in Plasma**
Edited by I.I. Sobelman

Volume 191 **Polarization of the Vacuum and a Quantum Relativistic Gas ...**
A.Ye. Shabad

Volume 190 (Supp. Vol.) **Superconductors with A15 Lattice and Bridge Contacts Based ...**
Edited by M. M. Sushchinskiy

Volume 190 **Metal-Optics and Superconductivity**
Edited by A.I. Golovashkin

Volume 189 **Clusters of Galaxies and Extragalactic Radio Sources**
Edited by A.D. Kuz'min

Volume 188 **Integral Systems, Solid State Physics and Theory of Phase Transitions. Pt. 2. Symmetries and Algebraic Structures in Physics**
Edited by V.V. Dodonov and V.I. Man'ko

Volume 187 **Quantum Field Theory, Quantum Mechanics and Quantum Optics: Pt. 1. Symmetries and Algebraic Structures in Physics**
Edited by V.V. Dodonov and V.I. Man'ko

Volume 186 **Photoproduction of Pions on Nucleons and Nuclei**
Edited by A.A. Komar

Volume 185	**Injection Lasers in Optical Communication and Information Processing Systems** Edited by Yu. M. Popov
Volume 184	**Electron Processes in MIS-Structure Memories** Edited by A.F. Plotnikov
Volume 183	**Invariants and the Evolution of Nonstationary Quantum Systems** Edited by M.A. Markov
Volume 182	**Luminescence of Wideband Semiconductors** Edited by M.D. Galanin
Volume 181	**Pulsed Metal and Metal Halogenide Vapor Lasers** Edited by G.G. Petrash
Volume 180	**Inelastic Light Scattering in Crystals** Edited by M.M. Sushchinskiy
Volume 179 (Supp. Vol. 2)	**Electron-Excited Molecules in Nonequilibrium Plasma** Edited by N.N. Sobolev
Volume 179 (Supp. Vol. 1)	**Interaction of Ultrashort Pulses with Matter** Edited by M.D. Galanin
Volume 179	**X-Ray Plasma Spectroscopy and the Properties of Multiply-Charged Ions** Edited by I.I. Sobel'man
Volume 178	**The Delfin Laser Thermonuclear Installation: Operational Complex ...** Edited by G.V. Sklizkov
Volume 177	**Stoichiometry in Crystal Compounds and Its Influence on Their Physical Properties** Edited by Yu. M. Popov
Volume 176 (Supp. Vol)	**Quantization, Gravitation and Group Methods in Physics** Edited by A.A. Komar
Volume 176	**Classical and Quantum Effects in Electrodynamics** Edited by A.A. Komar
Volume 175 (Supp. Vol.)	**Luminescence and Anisotropy of Zinc Sulfide Crystals** Edited by M.D. Galanin
Volume 175	**Luminescence Centers of Rare Earth Ions in Crystal Phosphors** Edited by M.D. Galanin
Volume 174 (Supp. Vol. 2)	**Electron Liquid Theory of Normal Metals** Edited by V.P. Silin
Volume 174 (Supp. Vol. 1)	**Nonequilibrium Superconductivity** Edited by V.L.Ginzburg
Volume 174	**Thermodynamics and Electrodynamics of Superconductors** Edited by V.L.Ginzburg
Volume 173	**Quantum Mechanics and Statistical Methods** Edited by M.M. Sushchinskiy
Volume 172	**Phase Conjugation of Laser Emission** Edited by N.G. Basov
Volume 171	**Research on Laser Theory** Edited by A.N. Orayevskiy
Volume 170	**The Theory of Target Compression by Longwave Laser Emission** Edited by G.V. Sklizkov
Volume 169	**The Physical Effects in the Gravitational Field Black Holes** Edited by M.A. Markov
Volume 168	**Issues in Intense-Field Quantum Electrodynamics** Edited by V.L. Ginzburg
Volume 167	**Group Theory, Gravitation and Elementary Particle Physics** Edited by A.A. Komar
Volume 166	**The Nonlinear Optics of Semiconductor Lasers** Edited by N.G. Basov
Volume 165	**Solitons and Instantons, Operator Quantization** Edited by V.L. Ginzburg

Horizons in World Physics
Volume 239

Theoretical Physics 2002, Part 1

Thomas F. George and Henk F. Arnoldus
Editors

Nova Science Publishers, Inc
New York

Senior Editors: Susan Boriotti and Donna Dennis
Coordinating Editors: Tatiana Shohov and Jennifer Vogt
Office Manager: Annette Hellinger
Graphics: Wanda Serrano
Editorial Production: Matthew Kozlowski, Jonathan Rose, and Maya Columbus
Circulation: Ave Maria Gonzalez and Vera Popovic
Communications and Acquisitions: Serge P. Shohov
Marketing: Cathy DeGregory

Library of Congress Cataloging-in-Publication Data
Available upon request.
ISBN 1-59033-435-3

400 Oser Ave, Suite 1600
Hauppauge, New York 11788-3619
Tele. 631-231-7269 Fax 631-231-8175
e-mail: Novascience@earthlink.net
Web Site: http://www.novapublishers.com

Printed in the United States of America

CONTENTS

PREFACE

The chapters in this book correspond to a selected set of articles published in 2002 in Volume 8 of the *International Journal of Theoretical Physics, Group Theory and Nonlinear Optics*. These chapters provide a representative sampling of the latest advances in theoretical physics.

Chapters 1 and 2 deal with the Hydrogen atom. In Chapter 1, Blaive and Cadilhac carry out an analysis of hydrogenoid atomic wave functions. In Chapter 2, Boudet, Blaive Geniyes and Vanel carry out a relativistic calculation with retardation of the photoelectric effect of Hydrogen.

Chapters 3 and 4 look at atoms in the presence of an external radiation field. Chapter 3 by Dastidar and Dastidar examines above-threshold ionization of Argon in a laser field. In Chapter 4, Kazakov applies the Jaynes-Cummings model to an atom interacting simultaneously with a quasiresonant classical field and a quantized mode.

Quantum dynamical problems are addressed in Chapters 5 and 6. In Chapter 5, Baute, Egusquiza and Muga study the effect of negative and classically-forbidden momenta in one-dimensional quantum scattering. Chapter 6 by Bellini finds analytical solutions to reaction-diffusion equations by mapping on a time-independent Schrödinger equation.

Chapters 7 and 8 are devoted to nuclear and particle physics. In Chapter 7, Kravchenko and Soznik obtain the nucleon-nucleus optical potential in the nuclear matter approximationw with the generalized Skyrme interation. In Chapter 8, Terasaki examines non-factorizable contributions in β decays.

The final three chapters contain various mathematical studies which are of interest to theoretical physics in general. In Chapter 9, Shiqing analyzes the equations of motion for the Newtonian n-body problem. Riazi looks at the geometry and topology of solitons in Chapter 10, and the book concludes with Chapter 11 containing a study by Elipe of the rotations of perturbed triaxial rigid bodies.

The Editors
2002

Theoretical Physics 2002, Part 1
Edited by Thomas F. George and Henk F. Arnoldus, pp. 1-10

ISBN 1-59033-435-3

Expansion of the Hydrogenoid Atomic Wave Functions in $1/n$. II. Besselian Expansion of Kummer Functions, and of the Atomic Wave Functions

Bruno Blaive* and Michel Cadilhac*

Abstract

The radial hydrogenlike atomic wave functions, which involve a Legendre polynomial (discrete spectrum) or a Kummer function $\Phi(\alpha, \gamma, 2r/a/(\gamma/2 - \alpha))$, are expanded into the family of besselian functions $B_n(\eta, z) = (z/\eta)^n J_n[2(\eta z)^{1/2}]$ studied in a former paper (Part I). As a preliminary result, a formula is given to expand any confluent hypergeometric function. The expansions obtained by Tricomi, for specific products of exponentials by Kummer functions, are revisited.

1. Introduction

The hydrogenlike atomic wave functions involve, in their radial part, a Legendre polynomial (discrete spectrum), or a Kummer function which is an infinite Taylor series (continuous spectrum). Since they are regular (holomorphic) functions, they can be expanded into the family of besselian functions

$$B_n(\eta, z) = (z/\eta)^{n/2} J_n[2(\eta z)^{1/2}], \tag{1}$$

whose properties were studied in a former paper (Part I) [1]. This will lead to the obtaining of a series expansion of the atomic wave functions into powers of the inverse principal quantum number n^{-1}.

For any regular function $f(z)$, two types of expansions into the functions (1) have been derived (Part I): (i) a unique *pure* expansion

$$f(z) = \sum_{n=0}^{\infty} b_n(\eta) B_n(\eta, z) \tag{2}$$

beginning from $n_0 = 0$ (or $n_0 > 0$), and (ii) an infinity of *impure* expansions

$$f(z) = \sum_{n=n_0}^{\infty} c_n(\eta) B_n(\eta, z) \tag{3}$$

*CNRS, Faculty of St-Jerome, 13397 Marseille Cedex 20, France E-mail: blaive@u3pic105.u-3mrs.fr

beginning from $n_0 < 0$. Both types will be considered in the present article. As a preliminary result, we shall apply expansion (2) to functions of the general form $z^r e^{-hz}\Phi(\alpha,\gamma,z)$, with independent parameters r,h,α,γ. Then, we shall treat two particular cases, the case $r=\gamma-1$ studied by Tricomi, and the case $r=0$ in which the coefficients b_n of Eq. (2) or c_n of Eq. (3) are given by simple recursion formulae. Finally, these recursion formulae will be applied to the case of the atomic wave functions.

2. Expansion of Kummer Functions and Related Functions

PROPOSITION. *We consider the function*

$$f(z) = z^r e^{-hz}\Phi(\alpha,\gamma,z), \tag{4}$$

involving a confluent hypergeometric function Φ *(cf. Glossary), with* $(\alpha,\gamma,h)\in\mathbf{C}^3$, γ *not a negative integer;* r *integer* ≥ 0. *Let* $\eta\in\mathbf{C}$.

Then, f *has the expansion of the form (2)*

$$f(z) = \sum_{n=r}^{\infty} b_n(\eta)B_n(\eta,z), \tag{5}$$

with

$$b_n(\eta) = \frac{\Gamma(\gamma)}{\Gamma(\alpha)}\sum_{q=0}^{n-r}\eta^{n-r-q}\frac{(q+r)!}{(n-r-q)!}\sum_{k=0}^{q}\frac{(-h)^{q-k}}{k!(q-k)!}\frac{\Gamma(\alpha+k)}{\Gamma(\gamma+k)}\quad(n\geq r), \tag{6}$$

and with the convention made for the Gamma functions (Glossary).

For $h=1$, *and when* α *and* γ *are non integral real numbers, whose difference* $\gamma-\alpha$ *is integral and* >1, *we have*

$$b_n(\eta) = \frac{\Gamma(\gamma)}{\Gamma(\gamma-\alpha)}\sum_{q=0}^{n-r}(-1)^q\frac{\eta^{n-r-q}(r+q)!}{q!(n-r-q)!}\frac{\Gamma(\gamma-\alpha+q)}{\Gamma(\gamma+q)}. \tag{7}$$

Proof. According to Eq. (A5) (Appendix), for $|p+h|>1$ the Laplace transform of f is

$$(\mathcal{L}f)(p) = [\mathcal{L}(x^r\Phi)](p+h) \tag{8}$$

$$= \frac{\Gamma(\gamma)}{\Gamma(\alpha)}\sum_{k=0}^{\infty}\frac{C_k}{(p+h)^{r+1+k}}, \tag{9}$$

where the coefficients C_k are defined by Eq. (A4). f is of exponential type, and we can use the Eq. (57) of Part I, to find the coefficients b_n of the expansion (2) of f:

$$\sum_{n=0}^{\infty}\frac{b_n}{p^{n+1}} = e^{\eta/p}\frac{\Gamma(\gamma)}{\Gamma(\alpha)}\sum_{k=0}^{\infty}\frac{C_k}{p^{r+1+k}}(1+\frac{h}{p})^{-r-1-k} \tag{10}$$

$$= \frac{\Gamma(\gamma)}{\Gamma(\alpha)} \left(\sum_{q=0}^{\infty} \frac{\eta^q}{q! p^q} \right) \sum_{k=0}^{\infty} \frac{C_k}{p^{k+r+1}} \sum_{l=0}^{\infty} \frac{h^l}{l! p^l} (-r-k-l) \ldots (-r-k-1) \tag{11}$$

for $|p| > |h|$, i.e.

$$\sum_{n=0}^{\infty} \frac{b_n}{p^{n+1}} = \frac{\Gamma(\gamma)}{\Gamma(\alpha)} \sum_{n=0}^{\infty} \frac{1}{p^{n+1}} \sum_{q=0}^{n-r} \frac{\eta^q (n-q)!}{q!} \sum_{k=0}^{n-r-q} \frac{\Gamma(\alpha+k)}{\Gamma(\gamma+k) k!} \frac{(-h)^{n-r-q-k}}{(n-r-q-k)!}. \tag{12}$$

Identification of the powers of $1/p$ yields $b_n = 0$ for $n < r$ and $b_n = (6)$ for $n \geq r$. For $h = 1$, the sum over k in (6) can be computed with the Vandermonde theorem, and this gives Eq. (7). Note that the hypothesis $h = 1$ is not restrictive for physical applications, since the variable z can be renormalized.

PROPOSITION. *Let $(\alpha, h) \in \mathbf{C}^2$, and γ be a positive integer. Then the function*

$$f(z) = e^{-hz} \Phi(\alpha, \gamma, z) \tag{13}$$

can be expanded into the form (3) (expansion beginning at $n_0 < 0$):,

$$f(z) = \sum_{n=1-\gamma}^{\infty} c_n(\eta) B_n(\eta, z), \tag{14}$$

where the coefficients c_n are given by the recursion formula (15):

$$c_{1-\gamma} = C, \qquad c_{2-\gamma} = C[\eta + (\gamma - 2)h - \beta],$$

$$(n+\gamma)c_{n+1} = c_n[\eta - h\gamma + \alpha - n(2h-1)] + c_{n-1}[\eta(2h-1) - nh(h-1)] + c_{n-2}\eta h(h-1) \quad (\textit{for } n \geq 3-\gamma). \tag{15}$$

We have introduced the constants

$$C(\alpha, \gamma) = (-1)^{\gamma-1} \frac{\Gamma(\gamma)\Gamma(-\alpha+1)}{\Gamma(-\alpha+\gamma)} \quad \textit{and} \quad \beta = -\alpha + \gamma - 1. \tag{16}$$

Proof. In the Laplace transform

$$(\mathcal{L}f)(p) = (\mathcal{L}\Phi)(p+h), \tag{17}$$

we substitute the expression (A2) for $\mathcal{L}\Phi$, and obtain

$$(\mathcal{L}f)(p) = C(p+h)^{\gamma-2} \left[D(p+h) - \sum_{q=0}^{\gamma-2} \frac{\Gamma(-\alpha+\gamma)}{\Gamma(-\alpha+\gamma-q) q!} \left(\frac{-1}{p+h} \right)^q \right], \tag{18}$$

where D is defined in (A12):

$$D(p+h) = \exp \left[\beta \ln \left(1 - \frac{1}{p+h} \right) \right]. \tag{19}$$

To obtain the coefficients c_n, we shall put Eq. (18) into the form (87) of Part I, choosing the last term $C(p+h)^{\gamma-2}\sum$ of Eq. (18) as the polynomial $-P(p)$ (of degree $\gamma - 2 = -n_0 - 1$):

$$(\mathcal{L}f)(p) - P(p) = Cp^{\gamma-2}\exp\left[(\alpha-1)\ln\left(1+\frac{h}{p}\right) + \beta\ln\left(1+\frac{h-1}{p}\right)\right]. \tag{20}$$

Introducing $t \equiv 1/p$ and the new coefficients

$$e_n = c_{n+1-\gamma}/C, \tag{21}$$

and derivating the two sides of Eq. (22),

$$\sum_{n=0}^{\infty} e_n t^n = \exp[\eta t + (\alpha-1)\ln(1+ht) + \beta\ln(1+(h-1)t)], \tag{22}$$

we obtain

$$\left(\sum_{n=1}^{\infty} e_n n t^{n-1}\right)[1+(2h-1)t+h(h-1)t^2]$$

$$= \left(\sum_{n=0}^{\infty} e_n t^n\right)\{[\eta+(\alpha-1)h+\beta(h-1)]+t[\eta(2h-1)+(\gamma-2)h(h-1)]+t^2\eta h(h-1)\}. \tag{23}$$

Identification of the coefficients of t^k on the two sides of this equation gives a recursion formula for the e_n's, which can be rearranged into Eq. (15).

COROLLARY (particular case $h = 0$). *Let $\alpha \in \mathbf{C}, \gamma$ integer > 0, $\eta \in \mathbf{C}$.*

Then, the Kummer function $\Phi(\alpha, \gamma, z)$ has the following expansion of the type (3):

$$\Phi(\alpha,\gamma,z) = C(\alpha,\gamma)\sum_{n=1-\gamma}^{\infty} c_n(\alpha,\gamma,\eta)B_n(\eta,z), \tag{24}$$

where $C(\alpha,\gamma)$ is defined by Eq. (16). The coefficients $c_n(\alpha,\gamma,\eta)$ are polynomials of degree $n-1+\gamma$ with respect to α,γ,η separately. They are given by the recursion formula

$$c_{1-\gamma} = 1 \quad c_{2-\gamma} = \eta - \beta \quad (\beta \equiv -\alpha+\gamma-1) \tag{25}$$

$$c_{n+1} = \frac{1}{n+\gamma}[(\eta+\alpha+n)c_n - \eta c_{n-1}] \quad (n \geq 2-\gamma). \tag{26}$$

The c_n's are also given by the explicit formulae

$$c_{1-\gamma} = 1 \tag{27}$$

$$c_n = \sum_{m=1}^{n-1+\gamma}\frac{1}{m!}\sum_J\prod_{i=1}^{m}A_{j_i}, \tag{28}$$

with

$$A_1 \equiv \eta - \beta, \quad A_k \equiv -\beta/k \ \ for \ \ k \geq 2; \tag{29}$$

J is a sequence, of variable length m ($m \geq 1$), of strictly positive integers $j_1, ..., j_m$ (some of them can be equal):

$$J = (j_1, ..., j_m), \quad with \ \ j_1 + j_2 + ... + j_m = n - 1 + \gamma. \tag{30}$$

Proof. The equation (26) is Eq. (15) written for $h = 0$. The explicit formulae (28), (29) were derived in ref.[2].

PROPOSITION. *Let $(\alpha, h) \in \mathbf{C}^2$, and γ be a positive integer. Then the function*

$$f(z) = z^{\gamma-1} e^{-hz} \Phi(\alpha, \gamma, z) \tag{31}$$

can be expanded into the form (2), and the coefficients b_n are given by the recursion formulae

$$b_n = 0 \ \ for \ \ n \lneq \gamma - 1 \ ; \ \ b_{\gamma-1} = (\gamma - 1)! \ ; \tag{32}$$

$$b_{\gamma+k}(k+1) = b_{k+\gamma-1}[\eta+\alpha-\gamma h-k(2h-1)]+b_{k+\gamma-2}[\eta(2h-1)-(\gamma+k-1)h(h-1)]+b_{k+\gamma-3}\eta h(h-1). \tag{33}$$

Proof. For $r = \gamma - 1$, equation (A5) simplifies into

$$\mathcal{L}(x^{\gamma-1}\Phi) = \frac{\Gamma(\gamma)}{\Gamma(\alpha)} \sum_{k=0}^{\infty} \frac{\Gamma(\alpha + k)}{k!} \frac{1}{p^{\gamma+k}}. \tag{34}$$

Using identity (A8), equation (34) can be rearranged into

$$\mathcal{L}(x^{\gamma-1}\Phi) = \frac{\Gamma(\gamma)}{p^\gamma} \exp\left[-\alpha \ln\left(1 - \frac{1}{p}\right)\right]. \tag{35}$$

We have

$$(\mathcal{L}f)(p) = [\mathcal{L}(x^{\gamma-1}\Phi)](p + h), \tag{36}$$

i.e.

$$e^{-\eta/p} \sum_{n=0}^{\infty} \frac{b_n(\eta)}{p^{n+1}} = \Gamma(\gamma)p^{-\gamma} \exp\left[(\alpha - \gamma)\ln\left(1 + \frac{h}{p}\right) - \alpha \ln\left(1 + \frac{h-1}{p}\right)\right]. \tag{37}$$

This equation is similar in form to Eq. (20), and equations (32), (33) can be derived in the same manner as Eq. (15).

Two particular cases of the recursion formula (33) were obtained by Tricomi: the case $h = \frac{1}{2}$, $\eta = \frac{\gamma}{2} - \alpha$,[4] and the case $\eta = -\alpha$ (arbitrary h).[5, 6]

3. The Kummer Function $\Phi(\alpha, \gamma, x/\zeta)$ Involved in the Atomic Wave Functions

The electronic quantum wave functions of hydrogen-like atoms involve an expansion of the type (3) (i.e. beginning from $n_0 < 0$) of the Kummer function $f(z) =$

$\Phi(\alpha, \gamma, z/\zeta)$ [2, 3]. In this physical case, the parameter η defining the expansion (3) is equal to 1. However, in order to introduce ζ into the hypergeometric function of Eq. (24), we shall choose $\eta = \zeta$ in this equation.

PROPOSITION. *Let $\alpha \in \mathbf{C}, \gamma$ integer > 0, ζ real > 0.*

Then the Kummer function $\Phi(\alpha, \gamma, z/\zeta)$ has the following expansion of the type (3):

$$\Phi(\alpha, \gamma, z/\zeta) = C(\alpha, \gamma) \sum_{k=0}^{\infty} \frac{c_k(\alpha, \gamma, \zeta)}{\zeta^k} B_{k+1-\gamma}(1, z), \tag{38}$$

with

$$c_k/\zeta^k = \sum_{q=0}^{k} \zeta^{-q} u_{q,k}(\alpha, \gamma). \tag{39}$$

$u_{q,k}$ is a polynomial of degree k with respect to α and γ separately.

In the particular case where $\zeta = -\alpha + \gamma/2$, we have

$$c_k/\zeta^k = \sum_{k/2 \le q \le k} \zeta^{-q} s_{q,k}(\gamma), \tag{40}$$

where the polynomials $s_{q,k}(\gamma)$, of degree $\le k$ with respect to γ, satisfy to the recursion formula

$$s_{1,1} = -\frac{\gamma}{2} + 1; \quad s_{1,2} = -\frac{1}{2} \tag{41}$$

$$(k+1) s_{q,k+1} = (k - \frac{\gamma}{2} + 1) s_{q-1,k} - s_{q-1,k-1} \quad (1 \le q \le k+1). \tag{42}$$

Proof. Let us change the variable z into z/η in expansion (24), and set $\eta = \zeta$. We obtain exactly Eq. (38), where the coefficients c_k are defined by Eq. (25), (26). These equations show that each c_k is a polynomial of degree k with respect to α, γ, ζ separately. Therefore c_k/ζ^k can be written in the form of Eq. (39).

The case when $\zeta = -\alpha + \gamma/2$: when α is replaced by $-\zeta + \gamma/2$ in $u_{q,k}(\alpha, \gamma)$, equation (39) can be written into the form

$$c_k/\zeta^k = \sum_{q=0}^{k} \zeta^{-q} s_{q,k}(\gamma). \tag{43}$$

In terms of the coefficients $s_{q,k}$, the recursion formula (26) becomes

$$(k+1) \sum_{q=0}^{k+1} \zeta^{k+1-q} s_{q,k+1} = (-\frac{\gamma}{2} + 1 + k) \sum_{q=1}^{k+1} \zeta^{k-q+1} s_{q-1,k} - \sum_{q=1}^{k} \zeta^{k-q+1} s_{q-1,k-1}. \tag{44}$$

Identification of the powers ζ^{k+1-q} in the two sides of Eq. (44) yields Eq. (42) for $1 \le q \le k$. This equation remains valid for $q = k+1$ if one agrees that $s_{q,k} = 0$ for $q > k$. We also obtain

$$s_{0,k} = 0, \quad \forall k. \tag{45}$$

Now, using induction on k, let us show that $s_{q,k} = 0$ for $q < k/2$. The property is verified for $k = 0$. Let us assume that it is true up to k, and establish it for $k+1$: we have

$$q < \frac{k+1}{2}, \tag{46}$$

hence

$$q - 1 < \frac{k-1}{2} < \frac{k}{2}. \tag{47}$$

Therefore, the rhs of Eq. (42) is zero, and $s_{q,k+1}$ is zero.

4. Conclusion

The hydrogen radial atomic wave functions R_{nl} involve a Kummer function $\phi(\alpha, \gamma, z)$ whose parameter α is either real ($\alpha = -n+l+1$ for the discrete spectrum), or pure imaginary ($\alpha = -in + l + 1$ for the continuous spectrum). Since α is any complex number in the expansion (38), this expansion can be applied to both spectra, discrete or continuous. More precisely, for the discrete spectrum, the radial wave function is

$$R_{nl}(r) = \frac{2}{(2l+1)!} n^{-3/2} \prod_{q=1}^{l} (n^2 - q^2)^{1/2} \left(\frac{2r}{n}\right)^l e^{-r/n} \Phi(-n+l+1, 2l+2, \frac{2r}{n}) \tag{48}$$

(the Bohr radius has been taken 1). And the same expression can be applied to the continuum, provided that n is replaced by in.

The substitution of expression (38) into (48), with the parameters

$$\alpha = -n + l + 1, \quad \gamma = 2l + 2, \quad \zeta = n, \quad z = 2r, \tag{49}$$

leads to

$$R_{nl}(r) = \frac{2}{n^{3/2}} \left[\prod_{k=1}^{l} \left(1 - \frac{k^2}{n^2}\right)^{-1/2} \right] e^{-x/(2n)}$$

$$\left[x^{-1/2} J_{2l+1} 2\sqrt{x} - \sum_{q=1}^{\infty} \frac{1}{n^q} \sum_{k=q}^{2q} s_{q,k}(l)\, x^{(k-1)/2} J_{k-2l-1} 2\sqrt{x} \right] \tag{50}$$

(with $x = 2r$), and Eq. (50) remains valid for the continuum when n is replaced by in. The polynomials $s_{q,k}(l)$, $q \le k \le 2q$ satisfy the recursion formulae (41), (42), i.e.

$$s_{1,1} = -l, \quad s_{1,2} = -\frac{1}{2}, \tag{51}$$

$$s_{q,k} = \left(1 - \frac{l+1}{k}\right) s_{q-1,k-1} - \frac{1}{k} s_{q-1,k-2}. \tag{52}$$

For the continuous spectrum, the rhs of Eq. (50) must be multiplied by a normalization factor,

$$e^{i\pi/4}(1 - e^{-2\pi n})^{-1/2}, \tag{53}$$

as described in ref. [3].

As examplified previously [2, 3], the series (50) converges quickly, so that a small number of terms are sufficient to obtain good approximations of R_{nl}, even for small values of the principal quantum number n.

Appendix: Laplace transform $\mathcal{L}\Phi$ of the confluent hypergeometric functions

PROPOSITION. *Let $(\alpha,\gamma)\in \mathbf{C}^2$, γ not a negative integer; $r\in \mathbf{N}$.*

Then, the function $x^r\Phi(\alpha,\gamma,x)$ has a Laplace transform for $|p|>1$,

$$\mathcal{L}(x^r\Phi)=\frac{\Gamma(r+1)}{p^{r+1}}F(\alpha,r+1,\gamma;1/p). \tag{A1}$$

When γ is a strictly positive integer, the Kummer function (case $r=0$) has the Laplace transform

$$(\mathcal{L}\Phi)(p)=(-1)^{\gamma-1}\frac{\Gamma(\gamma)\Gamma(-\alpha+1)}{\Gamma(-\alpha+\gamma)}p^{\gamma-2}$$

$$\left\{\exp\left[(-\alpha+\gamma-1)\ln\left(1-\frac{1}{p}\right)\right]-\sum_{q=0}^{\gamma-2}\frac{\Gamma(-\alpha+\gamma)}{\Gamma(-\alpha+\gamma-q)q!}\left(-\frac{1}{p}\right)^q\right\}, \tag{A2}$$

where ln *represents the principal determination of the complex logarithm (the imaginary part $>-\pi$ and $\leq\pi$).*

Proof. Using the definition of the Kummer function as a series, we see that the function

$$x^r\Phi(\alpha,\gamma,x)=\frac{\Gamma(\gamma)}{\Gamma(\alpha)}\sum_{k=0}^{\infty}C_k\frac{x^{k+r}}{\Gamma(r+1+k)}, \tag{A3}$$

where the coefficients C_k are defined by Eq. (A4)

$$C_k=\frac{\Gamma(\alpha+k)(k+r)!}{\Gamma(\gamma+k)k!}, \tag{A4}$$

is of exponential type. Its Laplace transform is

$$\mathcal{L}(x^r\Phi)=\frac{\Gamma(\gamma)}{\Gamma(\alpha)}\sum_{k=0}^{\infty}\frac{C_k}{p^{r+1+k}}. \tag{A5}$$

The p domain of convergence of the series (A5) is $|p|>1$. This is deduced easily from the Sterling formula, which gives an equivalent of the series coefficient when k tends to infinity:

$$C_k\sim k^{\alpha-\gamma+r}. \tag{A6}$$

Equation (A1), which is identical to (A5) was given by Tricomi [7].

For $r = 0$, and when γ is a positive integer, equation (A5) can be rewritten in the form

$$(\mathcal{L}\Phi)(p) = (-1)^{\gamma-1}\frac{\Gamma(\gamma)}{\Gamma(\alpha)}p^{\gamma-2}\sum_{k=0}^{\infty}\frac{(-1)^{\gamma}\Gamma(\alpha+k)}{\Gamma(\gamma+k)}\left(-\frac{1}{p}\right)^{\gamma+k-1}. \qquad (A7)$$

Using the identity

$$\Gamma(x)\Gamma(1-x) = (-1)^{x-y}\Gamma(y)\Gamma(1-y), \qquad (A8)$$

which is valid for complex numbers x and y whose difference $x-y$ is an integer, we obtain

$$(\mathcal{L}\Phi)(p) = (-1)^{\gamma-1}\frac{\Gamma(\gamma)\Gamma(-\alpha+1)}{\Gamma(-\alpha+\gamma)}p^{\gamma-2}\sum_{q=\gamma-1}^{\infty}\frac{\Gamma(-\alpha+\gamma)}{\Gamma(-\alpha+\gamma-q)q!}\left(-\frac{1}{p}\right)^{q}. \qquad (A9)$$

In Eq. (A9), $\sum$ can be replaced by

$$\sum_{q=\gamma-1}^{\infty} = \sum_{q=0}^{\infty} - \sum_{q=0}^{\gamma-2}, \qquad (A10)$$

with

$$D(p) \equiv \sum_{q=0}^{\infty}\frac{\Gamma(-\alpha+\gamma)}{\Gamma(-\alpha+\gamma-q)q!}\left(-\frac{1}{p}\right)^{q} \qquad (A11)$$

$$= \exp\left[(-\alpha+\gamma-1)\ln\left(1-\frac{1}{p}\right)\right]. \qquad (A12)$$

Equation (A12), which generalizes the Newton binomial formula to a nonreal power $-\alpha+\gamma-1$, can be verified in the following manner: if we consider expression (A12) as a function f(z) of the variable $z = 1-1/p$, its nth derivative gives the coefficients of the series (A11). This property, together with the limit $D(p) \to 1$ when $p \to +\infty$, shows that the series (A11) is the Taylor series of $f(z)$.

The substitution of (A10) and (A12) into (A9) gives Eq. (A2).

Glossary

$\Phi(a,b,z) = \frac{\Gamma(b)}{\Gamma(a)}\sum_{k=0}^{\infty}\frac{\Gamma(a+k)}{\Gamma(b+k)k!}z^k$ confluent hypergeometric or Kummer function ${}_1F_1$

$\mathcal{L}$ Laplace transform $(\mathcal{L}f)(p) = \int_0^{\infty} dx\, e^{-px}f(x)$ or analytic continuation

$\Gamma(z)$ Euler Gamma function. Convention: when α is a strictly negative integer and $k \in \mathbf{N}$, the fraction $\Gamma(\alpha+k)/\Gamma(\alpha)$ means $\alpha(\alpha+1)...(\alpha+k-1)$ for $k \le -\alpha$, and 0 for $k > -\alpha$.

References

[1] B. Blaive, M. Cadilhac, Expansion of the hydrogenoid wave functions. I. A besselian extension of the Taylor expansion, and its group structure, *Int. J. Theor. Phys., Group Theory, Nonlinear Optics*, 7(3), 57-73 (2001).

[2] B. Blaive, M. Cadilhac, Upper limit of the discrete hydrogen-like wave functions: Expansion in the inverse principal quantum number n^{-1}, *J. Math. Phys.*, 38, 6061-6071 (1997).

[3] B. Blaive, M. Cadilhac, Expansion of the hydrogen-like wave functions in the inverse principal quantum number, for the discrete and continuous spectra, *J. Math. Phys.*, 39, 5693-5699 (1998).

[4] F. G. Tricomi, *Funzioni ipergeometriche confluenti*, Eqs. (1) p. 37, (6) p. 39, (9) or (9') p. 41. Ed. Cremonese, Roma (1954).

[5] F. G. Tricomi, Sviluppo dei polinomi di Laguerre e di Hermite in serie di funzioni di Bessel, Eq. (3) p. 15, *Giorn. Ist. Ital. Attuari*, 12, 14-33 (1941).

[6] Ref. 4, Eq. (10) p. 42.

[7] Ref. 4, Eq. (7) p. 25.

Theoretical Physics 2002, Part 1
Edited by Thomas F. George and Henk F. Arnoldus, pp. 11-20
ISBN 1-59033-435-3

Numerical Relativistic Calculation with the Retardation of the Photoeffect of Hydrogenic Atoms

Roger Boudet*, Bruno Blaive†, Stephane Geniyes* and Michel Vanel*

Abstract

The aim of this paper is the numerical calculation with retardation, for a central potential, of transitions between the discrete states 1S1/2 and states of the continuum (photoeffect) $P1/2$ and $P3/2$, on ground of formulas previously established. Two kinds of formulas are used. The first ones, based on combinations of hypergeometric series, allow a calculation suitable for all value of Z. The second ones, based on the Laplace method, imply the approximation $Z^2\alpha^2 \ll \kappa^2$ and allow a convenient calculation and also a verification of the results for small values of Z. This paper is more precisely devoted to numerical results using the second method and their comparison with the ones obtained by the first method, and also the estimation of the corrections that the relativistic calculation brings to the retarded non relativistic one.

1 Introduction

In previous papers [1]-[3] the general relativistic formulas with and without retardation have been established for all the transitions $S1/2 - P1/2$ and $S1/2 - P3/2$ of the photoeffect for a central potential. In Ref. [2] and [3] these formulas have been considered for the transitions implying the two states $1S1/2$. In Ref. [3] they have been expressed as combinations of hypergeometric series, suitable for all value of Z, whose numerical calculation may easily be achieved on a computer up to any required degree of precision. They have been achieved also, as a verification, in the case of the approximation $Z^2\alpha^2 \ll \kappa^2$, by using the Laplace method leading to the integral representation in the complex plane of the confluent hypergeometric functions and the use of the residues theorem.

In Ref. [3] some numerical results have been calculated with the help of the first method and a numerical confirmation of the conformity of the relativistic retarded formulas with the non relativistic non retarded ones for values of the free energy close to mc^2 have been made.

*Université de Provence, Pl. V. Hugo, 13331 Marseille, France; e-mail: boudet a gyptis. univ-mrs.fr

†CNRS, Faculté des Sciences de St-Jérôme,13397 Marseille, France; e-mail: blaive a u3pic105.u-3mrs.fr

The aim of the paper is the presentation of the numerical results obtained by the second method and by the formulas of Ref. [2] of the relativistic, non retarded case. Furthermore, a numerical evaluation of the errors made by the use of the retarded non relativistic formulas with respect to the retarded relativistic ones have been achieved.

We recall that the general relativistic formulas with retardation of the transition between two states, established in Ref. [1], are based on the probability transition current $\mathbf{J}$ defined by

$$J^k(x^0, \mathbf{r}) = \overline{\Psi}_2 \gamma^k \Psi_1 + \overline{\Psi}_1 \gamma^k \Psi_2, \quad x^0 = ct \ \ (k = 1, 2, 3) \tag{1}$$

where Ψ_1, Ψ_2 are the Dirac wave functions corresponding to the levels of energy E_1, E_2. $\mathbf{J}$ is time-periodic:

$$\mathbf{J}(x^0, \mathbf{r}) = \cos \Omega x^0 \mathbf{j}_1(\mathbf{r}) + \sin \Omega x^0 \mathbf{j}_2(\mathbf{r}), \quad \Omega = (E_1 - E_2)/\hbar c \tag{2}$$

The matrix elements relative to the transition between the two states are defined by (see Ref. [4], Eq. (73.1))

$$D_j(\mathbf{k}, \mathbf{N}) = \frac{mc}{\hbar} \mathbf{T}_j^{\perp}(\mathbf{k}).\mathbf{N}, \quad (j = 1, 2) \tag{3}$$

with

$$\mathbf{T}_j^{\perp}(\mathbf{k}) = \frac{1}{2} \int e^{i(\mathbf{k}.\mathbf{r})} \, \mathbf{j}_j^{\perp}(\mathbf{r}) \, d\tau, \quad (d\tau = dx^1 dx^2 dx^3) \tag{4}$$

$\mathbf{k}$ is the propagation vector of the incident wave so that

$$k = |\mathbf{k}| = \frac{E_2 - E_1}{\hbar c} \tag{5}$$

$\mathbf{N}$ is an unit vector, orthogonal to $\mathbf{k}$, parallel to the polarization direction, $\mathbf{X}^{\perp}$ means the component orthogonal to $\mathbf{k}$ of a vector $\mathbf{X}$.

Here E_1 will be the energy of the $1S1/2$ states and $E = E_2$ the energy of a state $P1/2$ or $P3/2$ of the continuum.

2 The Expressions of the Vectors $\mathbf{T}_j^{\perp}(k)$

The expressions of the vectors $\mathbf{T}_j^{\perp}(\mathbf{k})$ have been established in Ref. [1] (Eqs. (28) to (37)) for all the degeneracies of the transitions. Here, we will consider only the cases where the variation in the transition Δm of the magnetic number is null.

We recall the results established in Ref. [1]. Denoting

$$L_{s,ij}(k) = \int_0^{\infty} j_s(kr) g_i(r) f_j(r) r^2 dr \tag{6}$$

where $g_i(r)$ (great) and $f_i(r)$ (fine) ($i = 1, 2$) are the radial components of the Darwin solutions of the Dirac equation for the central potential $eZ/4\pi\varepsilon_0 r$ of the states Ψ_1, Ψ_2 and where $j_s(x)$ represents the spherical Bessel functions such that $s = 0$ or $s = 2$, one has

1. *Transitions* $S1/2 - P1/2$ ($\kappa_1 = -1$, $\kappa_2 = 1$, $\Delta m = 0$):

$$\mathbf{T}_1^{\perp}(\mathbf{k}) = 0 \tag{7}$$

$$\mathbf{T}_2^{\perp}(\mathbf{k}) = \pm\frac{1}{3}[(-3L_{0,12} - L_{0,21} + 2L_{2,21})(k)]\sin\theta_0\,\mathbf{I} \tag{8}$$

2. *Transitions* $S1/2 - P3/2$ ($\kappa_1 = -1$, $\kappa_2 = -2$, $\Delta m = 0$):

$$\mathbf{T}_1^{\perp}(\mathbf{k}) = \pm\frac{3}{\sqrt{2}}[(L_{2,12} + L_{2,21})(k)]\cos\theta_0\sin\theta_0\,\mathbf{J} \tag{9}$$

$$\mathbf{T}_2^{\perp}(\mathbf{k}) = \frac{\sqrt{2}}{6}[(-4L_{0,21} + 3L_{2,12} - L_{2,21})(k)]\sin\theta_0\,\mathbf{I} \tag{10}$$

The angle θ_0 and the unit vector $\mathbf{I}$ are such that

$$\mathbf{e}_3 = \cos\theta_0\mathbf{K} + \sin\theta_0\mathbf{I}$$

where $\mathbf{e}_3$ is the unit vector defining the positive x^3 axis of the Darwin solution, $\mathbf{K} = \mathbf{k}/k$ and $(\mathbf{I}, \mathbf{J}, \mathbf{K})$ is a positive orthonormal frame.

Introducing the dimensionless numbers

$$\rho = \frac{Zr}{a_0}, \qquad K = \frac{ka_0}{Z}$$

where $a_0 = \hbar/\alpha mc$ is the Bohr radius, and the parameters

$$\gamma_1 = [1 - \alpha^2 Z^2]^{1/2}, \ \delta = \left[1 + \frac{\alpha^2 Z^2}{\gamma_1^2}\right]^{-1/2}$$

$$C = 2^{\gamma_1 - 1}\left[\frac{1+\delta}{\Gamma(2\gamma_1 + 1)}\right]^{1/2}, \quad d = \frac{2}{\alpha Z}\left[\frac{1-\delta}{1+\delta}\right]^{1/2}$$

associated with the state $\Psi_1 = 1S1/2$ (see Ref. 4, Eq. (14.38)), then

$$\gamma = [\kappa^2 - \alpha^2 Z^2]^{1/2}, \ \varepsilon = \frac{E}{mc^2}, \ n = \frac{\alpha Z}{[\varepsilon^2 - 1]^{1/2}}, \ \nu = [n^2 + \alpha^2 Z^2]^{1/2} \tag{11}$$

associated with a state $\Psi_2 = P1/2$ ($\kappa = 1$) or $P3/2$ ($\kappa = -2$) of energy E of the continuum, one obtains (see Ref. [3], Eqs. (13)-(22))

$$L_{s,12}(k) = \frac{2}{\sqrt{n}[\nu + n]^{1/2}}\mathrm{Im}[H(n)I_s(K,n)] \tag{12}$$

$$L_{s,21}(k) = -d\frac{[\nu + n]^{1/2}}{\sqrt{n}}\mathrm{Re}[H(n)I_s(K,n)] \tag{13}$$

where

$$H(n) = C\frac{e^{\nu\pi/2}|\Gamma(\gamma + 1 - i\nu)|\sqrt{n}}{\sqrt{\pi}\,\Gamma(2\gamma + 1)}\left[\frac{2}{n}\right]^{\gamma}\left[\frac{-\kappa - in}{\gamma + i\nu}\right]^{1/2} \tag{14}$$

$$I_s(K,n) = \int_0^{\infty}\exp[-(1 + \frac{i}{n})\rho]j_s(K\rho)\rho^{\gamma_1+\gamma}F(\gamma + i\nu; 2\gamma + 1; i\frac{2\rho}{n})d\rho \tag{15}$$

3 The Two Methods of Calculation of the Integrals $L_{s,ij}(k)$

The problem of the determination of the vectors $\mathbf{T}_j^{\perp}(k)$ is reduced to the calculation of the integrals (15), i.e. integrals of the form

$$I_s(K,n) = \int_0^{\infty} e^{-\lambda\rho}\rho^b j_s(K\rho)F(a;c;\mu\rho)d\rho \tag{16}$$

with

$$j_0(x) = \frac{\sin x}{x}, \quad j_2(x) = 3\left[\frac{\sin x}{x^3} - \frac{\cos x}{x^2}\right] - \frac{\sin x}{x} \tag{17}$$

where $\lambda = 1 + i/n$, $a = \gamma + i\nu$, $b = \gamma_1 + \gamma$, $c = 2\gamma + 1$, $\mu = i2/n$.
Two methods have been used in Ref. [3].

1. The first one is based on an integration term by term and leads to a combination of hypergeometric functions and thus to the calculation of series.
Defining

$$\lambda_{\pm} = \lambda \pm iK, \tag{18}$$

one obtains (see Ref. [3], Eqs. (25)-(33))

$$I_0(K,n) = \frac{\Gamma(b)}{i2K}\left[\frac{1}{\lambda_-^b}F(a,b;c;\frac{\mu}{\lambda_-}) - \frac{1}{\lambda_+^b}F(a,b;c;\frac{\mu}{\lambda_+})\right], \quad b > 0 \tag{19}$$

$$M_1(K,n) = \frac{\Gamma(b-1)}{2}\left[\frac{1}{\lambda_-^{b-1}}F(a,b-1;c;\frac{\mu}{\lambda_-}) + \frac{1}{\lambda_+^{b-1}}F(a,b-1;c;\frac{\mu}{\lambda_+})\right], \quad b > 1 \tag{20}$$

$$M_2(K,n) = \frac{\Gamma(b-2)}{i2K}\left[\frac{1}{\lambda_-^{b-2}}F(a,b-2;c;\frac{\mu}{\lambda_-}) - \frac{1}{\lambda_+^{b-2}}F(a,b-2;c;\frac{\mu}{\lambda_+})\right], \quad b > 2 \tag{21}$$

and

$$I_2(K,n) = \frac{3}{K^2}[M_2(K,n) - M_1(K,n)] - I_0(K,n) \tag{22}$$

In the numerical calculation, the hypergeometric *functions* $F(a,\beta;c;z)$ are to be converted into hypergeometric *series*, easily calculable on a computer. In the case of $P1/2$, Eq. (21) cannot be directly applied to the calculation of $M_2(K,n)$ because $b = \gamma_1 + \gamma = 2 - \xi$, with $0 < \xi < 1$. A suitable decomposition of $M_2(K,n)$ allows its calculation (see Ref. [3], Eqs. (31)-(33)).

2. The second way uses the Laplace method and the representation in the complex plane of the confluent hypergeometric functions and leads to a combination of usual functions. In fact the approximation $Z^2\alpha^2 \ll \kappa^2$ allows one to replace in Eq. (16) Re(a), b, c by integers and to use the residues theorem.

This method is suitable only for small values of Z, but allows a verification of the validity of the results obtained by the first method. Furthermore it is the natural continuation of an approach initiated in Ref. [5] and which has been used for a long time in the study of the photoeffect (Ref. [6], [4]).

Denoting

$$\Theta = \frac{1}{2}\left(\cot^{-1}\left[\frac{n}{nK+1}\right] - \cot^{-1}\left[\frac{n}{nK-1}\right]\right), \quad \phi = \frac{1}{2}\ln\left[\frac{n^2+(nK+1)^2}{n^2+(nK-1)^2}\right]$$

$$u = \cos\nu\Phi, \quad w = \frac{\sin\nu\Phi}{2nK} \tag{23}$$

$$U = \frac{\nu(u-2(n+i)w)}{[n(i+K)-1][(n(i-K)-1]}, \quad V = \frac{(n-i)}{n^2K^2}(\nu u - (n^2+i)w)$$

one obtains (see Ref. [3] Eqs. (34)-(48))

(i) For $P1/2$ ($b=2$, $a=1+i\nu$, $c=3$)

$$D = \frac{1}{2(1-i\nu)(-i\nu)}, \quad R_h = -ie^{-2\nu\Theta}(u+U) \tag{24}$$

$$I_0(K,n) = D2n^3(-R_h) \tag{25}$$

Note that the calculation of $I_2(K,n)$ cannot be achieved without complications by this method.

(ii) For $P3/2$ ($b=3$, $a=2+i\nu$, $c=5$)

$$D = \frac{3}{2i\nu(2-i\nu)(1+\nu^2)} \tag{26}$$

$$R_f = \frac{e^{-2\nu\Theta}}{4}[4u + (10\nu+9n-3i)w + 2(\nu-i)U - V] - 1 \tag{27}$$

$$R_g = \frac{e^{-2\nu\Theta}}{4n^2K^2}[(2\nu-3n+i)u + 2(\nu-i)U + 3V] \tag{28}$$

$$I_0(K,n) = D[6n^4(-R_f) - 2n^6K^2(-R_g)] \tag{29}$$

$$I_2(K,n) = -D8n^6K^2(-R_g) \tag{30}$$

4 Some Numerical Results

The numerical results presented here concern the square of the matrix elements deduced from Eqs. (7), (8) for $P1/2$ and (9), (10) for $P3/2$, in which $\theta_0 = \pi/2$, i.e. the number

$$L(k) = |\mathbf{T}_{\frac{1}{2}}^{\perp}(\mathbf{k})|^2, \quad \text{where} \quad k = |\mathbf{k}| = \frac{E-E_1}{\hbar c} = \frac{2\pi}{\lambda}, \quad \mathbf{k}.\mathbf{e}_3 = 0 \tag{31}$$

E_1 is the energy of $1S1/2$ and E the energy of a state $P1/2$ or $P3/2$ of the continuum, λ the wave length of the incident wave.

The number $L(k)$ is dimensionless but to be in conformity with the definition of Ref. [4], Eq. (73.1), it may be considered as a quantity expressed in unit $(mc^2/\hbar)^2 = (1/\alpha a_0)^2$.

The calculation gives results obtained by the Laplace method and their comparison with the ones obtained by the use of the hypergeometric functions.

In the limit $k \to 0$, Eqs. (41)-(48) of Ref. [3] allow one to recover the expressions of the numbers $L(0) = |\mathbf{T}\frac{1}{2}(0)|^2$ (see Ref. [2], Eqs. (5), (44), (52)) corresponding to the relativistic, non retarded case.

In the Pauli approximation (here $\nu = n$ in addition to $Z^2\alpha^2 \ll \kappa^2$) the numbers $L(0)$ satisfy the relations $L(0) = |X_{E1}|^2/3$ and $L(0) = 2|X_{E1}|^2/3$ for the transitions S1/2-P1/2 and S1/2-P3/2 respectively, where $|X_{E1}|^2$ (see Ref. [4], Eq. (71.4)) is the matrix element of a transition $1s - p_0$ in the Schrödinger theory. (We recall that this last number has been used for a long time as the main element of the theoretical verification of the experiments on the photoeffect (Ref. [4], Sect. 71)). This property has been analytically verified on the relativistic non retarded formulas of Ref. [2], and numerically verified for the small values of k on the relativistic retarded formulas of Ref. [3].

Here we compare $L(k)$ with the non relativistic retarded matrix elements of the transitions $1s - p_0$ established in conformity with the calculation of Sommerfeld and Schur (see Eq. (65) of Ref. [2] with $\theta_0 = \pi/2$), multiplied by 1/3 and 2/3. That gives the error made by the use of the non relativistic retarded theory.

1. In Tables I, II, III, the line (a) gives the values $L_e(k)$ of $L(k)$ calculated with the use of hypergeometric functions (Eqs. (16)-(18), (20)-(22)).

We emphasize that the high degree of precision allowed by this method, concerns only atoms considered strictly as hydrogenic and whose number Z in not too large, in such a way that the corrections due to the presence of other electrons or the size of the nucleus are not to be taken into account.

For simplicity, because the calculation of $I_2(K, n)$ for the transitions 1S1/2-P1/2 requires sophisticated numerical methods, we did not calculate this number. For the values of $K = a_0 k/Z$ not too large its contribution is negligible. For the large values of K the matrix elements of these transitions are not mentioned.

$\lambda(\mathring{A})$	$E - mc^2$ (Kev)	$K = a_0 k/Z$		$1S1/2 - P1/2$	$1S1/2 - P3/2$
		Table I:	H	$(Z = 1)$	
828.42	0.00136	0.004	a	0.110973	0.221954
			b	1.00005	1.00007
			c	1.00004	1.00003
			d	1.00005	1.00001
0.608	20.4	5.47	a	1.0997×10^{-8}	2.3179×10^{-8}
			b	0.99988	0.99982
			c	0.9357	1.0034
			d	1.1080	1.0514
0.0091	1360.6	364.9	a	3.37×10^{-13}	6.035×10^{-13}
			b	0.99963	1.0021
			c	0.1479	0.3276
			d	0.9434	1.0553
0.006	2040.8	547.3	a	–	3.032×10^{-13}
			b	–	1.0024
			c	–	0.1625
			d	–	0.093
9×10^{-5}	136056	36487	a	–	9.40×10^{-16}
			b	–	1.001
			c	–	4.83×10^{-6}
			d	–	2.44×10^{-11}
		Table II:	Na	$(Z = 11)$	
7.457	0.0165	0.041	a	0.12723	0.25549
			b	1.006	1.009
			c	1.005	1.003
			d	1.006	1.002
0.502	23.05	0.602	a	5.692×10^{-4}	1.2282×10^{-3}
			b	1.0004	0.995
			c	0.927	1.006
			d	1.141	1.057
0.075	162	4.01	a	3.73×10^{-6}	1.3317×10^{-5}
			c	0.585	0.997
			d	2.6	1.45
0.015	821	20	a	–	2.283×10^{-7}
			c	–	0.611
			d	–	4.8

$\lambda(\AA)$	$E - mc^2$ (Kev)	$K = a_0k/Z$		$1S1/2 - P1/2$	$1S1/3 - P3/2$
		Table III:	Cs	$(Z = 55)$	
0.3	0.041	0.2	a	0.10806	0.24347
			b	1.170	1.251
			c	1.139	1.070
			d	1.164	1.033
0.2	20.57	0.301	a	0.05	0.1226
			c	1.067	1.071
			d	1.301	1.038
0.03	370	2	a	–	0.00219
			c	–	0.955
			d	–	1.8

Tables I-II-III : Matrix elements $L(k)$ (Eq. (31).

$L_e(k)$ (hypergeometric functions), $L_a(k)$ (Laplace method), $L(0)$ relativistic non retarded, $L_0(k)$, non relativistic retarded.

$a = L_e(k), b = L_a(k)/L_e(k), c = L(0)/L_e(k), d = L_0(k)/L_e(k)$.

The line (b) gives the ratios $L_a(k)/L_e(k)$ where the numbers $L_a(k)$ are obtained by the Laplace method (P1/2: Eq. (25); P3/2: Eqs. (29)-(30)) and allows a numerical comparison of the two methods.

The line (c) gives the ratios $L(0)/L_e(k)$ where the numbers $L(0)$ correspond to the relativistic non retarded cases ($k = 0$) (P1/2: Ref. [2], Eq. (44); P3/2: Eq. (19)) and allows the evaluation of the incidence of the retardation in the relativistic calculation.

The line (d) gives the ratios $L_0(k)/L_e(k)$ where the numbers $L_0(k)$ correspond to the non relativistic retarded cases (see Ref. [2], Eq. (65)) with $\theta_0 = \pi/2$ and a factor 1/3 for P1/2, 2/3 for P3/2). It allows the estimation of the errors which are made when the retarded non relativistic calculation is used.

2. Comments

For E close to mc^2, i.e. for K small, and small values of Z, all the lines (b) to (d) must be close to 1, as can be seen on Table I ($Z = 1$) for K=0.004. That constitutes a very credible confirmation of the validity of all the formulas used in this paper.

For small values of Z the line (b) must be close to 1 as that can be seen on table I (Z=1) and II ($Z = 11$). So, for these values, the formulas (23)-(30) whose running time on a computer is shorter than for the ones of the hypergeometric series, can be used (for any value of K, for P2/3, and values of K not too large, for P1/2).

The incidence of the retardation, in the relativistic calculation, begins to be appreciable (for $Z = 1$) around $E - mc^2 = 300$ Kev and then becomes important, as it is shown in the line (c).

The errors (line (d)) due to the use of the retarded non relativistic formulas, with

respect to the retarded relativistic ones, are not very large for $E - mc^2 < 1300$ Kev ($Z = 1$), but then they increase fast, and become considerable for E very large. That shows the profound difference between the relativistic and the non relativistic calculations. One can notice on table II and III the incidence of the value of Z on these errors. We recall that the non relativistic calculation cannot be applied in the cases of degeneracies (for reasons analog to the differences between the normal and anomal Zeeman effect), as it is shown by the formulas (26)-(37) of Ref. [1].
As an additional verification, we have add the comparison of the two methods for the calculation of the *same* integrals $I_s(K, n)$, Eq. (16), by introducing the approximation $Z^2\alpha^2 \ll \kappa^2$ in the first method. One can notice a large difference between the formulas (18)-(22) and (23)-(30) of the two methods. Nevertheless, if the number of terms taken into account in the calculation of the hypergeometric series is sufficiently large, one can obtain a numerical coincidence on the results reaching 16 numerals.
For example, for $P3/2$ ($\kappa = -2$), $Z = 1$, and so in Eq. (16), $b = 3$, $a = 2 + i\nu$, $c = 5$, $\nu = [n^2 + \alpha^2]^{1/2}$, one obtains for one million terms, the *same* number by using the *two* methods:
$I_2(100, 1) = 0.263884625639136 - 0.001954355294650i$
But with a number of terms *less than* 20, one obtains with the first method
$I_2(100, 1) = 0.263884625675413 - 0.001954355294619i$
i.e. a results exact on 11 numerals. The very good convergence of the hypergeometric series in this particular case have been verified for the other values of K and n.

5 Conclusion

The present paper has confirmed numerically the validity of the formulas giving the exact relativistic matrix elements with retardation of the photoeffect of hydrogenic atoms, by the concordance of the results in the two methods used. Furthermore it has shown the necessity of the use of the relativistic calculation in place of the non relativistic one, even when the retardation is taken into account in this last calculation.

References

[1] R. Boudet, *Found. of Phys.*, **29**, 29 (1999).

[2] R. Boudet, *Found. of Phys.*, **29**, 49 (1999).

[3] R. Boudet and B. Blaive *Found. of Phys.*, **30**, 1283 (2000).

[4] H. Bethe and E. Salpeter, *Quantum Mechanics of one and two-electron Atoms* (Springer, Berlin, 1957).

[5] A. Sommerfeld and G. Schur, *Ann. d. Phys.*, **4**, 409 (1930).

[6] H. Hall, *Rev. Mod. Phys.*, **8**, 358 (1936).

Theoretical Physics 2002, Part 1
Edited by Thomas F. George and Henk F. Arnoldus, pp. 21-25
ISBN 1-59033-435-3

On the Absence of Electron Correlation Effects in the Photoelectron Spectra for Above-Threshold Ionization of Atoms

T K Rai Dastidar*and Krishna Rai Dastidar †

Abstract

We point out that a recently observed, apparently perplexing feature of above-threshold ionization of Argon in a laser field can be explained very simply, while some closely related phenomena have been predicted earlier.

I. Introduction

In a recent work [1] on above-threshold ionization (ATI) of Argon in laser fields of intensity ranging over several tens of TW/cm^2, the authors obtain a feature that they find perplexing: in the photoelectron spectra they find no evidence of electron correlation, although double ionization is certainly taking place at these intensities.

The purpose of this note is to point out that this feature is not unexpected. Indeed, the absence of electron correlation in another feature of ATI, namely, the suppression or "quenching" of the lower-order peaks at high intensities, was already predicted on the basis of a non-local model of quantum electrodynamics (QED) quite a long time ago [2], where the model was proposed in an *ad hoc* manner; subsequently we placed the theory on a field-theoretic basis [3]. In both these papers the aspect of double ionisation in strong laser fields was also discussed in detail, with focus on the role of electron correlation on this phenomenon. However, before going into these new theoretical developments, let us discuss some generalities.

In a multi-electron atom exposed to strong radiation, electron correlation can be expected if and only if the electric field strength does not "overwhelm" the typical interelectronic Coulomb repulsion in the valence shell of the atom. According to the Wannier picture [4], the two outer-shell electrons in a closed-shell atom are most strongly correlated when they are at diametrically opposite positions with respect to the nucleus, i.e., when $\mathbf{r}_1 = -\mathbf{r}_2$. In atomic units, the Coulomb repulsion between the two electrons is $\frac{1}{4}r^{-2}$. In order for the signature of electron correlation to be visible in the MPI/ATI spectra of an atom placed in a laser, it is imperative that the electric field strength of the latter be significantly lower than this. For Argon, the outer-shell radius is $\sim$ 10 a.u.; thus the above field strength F

* This work is dedicated to his memory

† Indian Association for the Cultivation of Science, Calcutta 700 032, India. Electronic address : spkrd@mahendra.iacs.res.in

(say) is $\sim 2.5 \times 10^{-3}$ a.u. A laser electric field F corresponds to an energy density of $F^2/4\pi$ a.u., i.e., $\sim 5 \times 10^{-7}$ hartree a_0^{-3}, where a_0 is the atomic unit of length. Multiplying by c ($\simeq 137\ a_0/t_0$, t_0 being the atomic unit of time) gives a radiant energy flux of $\sim 7 \times 10^{-5}$ hartree $a_0^{-2}t_0^{-1}$, i.e. $\sim 4.5 \times 10^{11}$ W/cm^2. The laser intensities used in [1] are two orders of magnitude larger than this ; as such, signatures of electron correlation cannot be expected in the photoelectron spectra.

As regards *double ionisation*, it should be noted that this can take place both *sequentially* as well as *directly*. The former does not depend on electron correlation, while the latter does. The effect of electron correlation on this direct double ionisation (DDI)[a] has been demonstrated in [2] as well as in [3], where we also showed that another, frequently accompanying feature found in ATI, namely the suppression or "quenching" of lower-order ATI peaks at high intensities, *would not be affected by electron correlation*. With the increase in laser intensity, electron correlation will break down, leading to damping of DDI processes. Laser intensity dependence of DDI considering this damping effect will be studied elsewhere.

The central result of this theory [2,3] was that the interaction of this nonlocal field with matter allowed the emission/absorption of *two phase-correlated photons* in the first order. This feature of linear intensity dependence of the two-photon absorption process has been observed [5] in the interaction of a squeezed vacuum with an atom. In what follows we very briefly outline the basic theory and show how the above features of ATI as found in [1] follow from it almost trivially.

II. Brief Theory

For the detailed theoretical development the reader is referred to [3] ; here we just give the main results. In this nonlocal model, the electron-radiation interaction matrix element in the first order is given by

$$M_{fi}(t) = \langle \phi_f(x) | \mathbf{p}.\mathcal{A}(x, x') | \phi_i(x') \rangle, \tag{1}$$

where $x \equiv (\vec{r}, t)$, $x' \equiv (\vec{r}', t')$, and the integration runs over d^4x' and d^3r. A Fourier expansion of $\mathcal{A}$, which is a function of two variables (see [6]), gives

$$\mathcal{A}(x, x') = \sum_{\mathbf{k}_1\lambda_1} \sum_{\mathbf{k}_2\lambda_2} [C_{12}\hat{\epsilon}_{12}(\hat{r}, \hat{r}') \exp(i\mathbf{k}_1.\mathbf{r} + i\mathbf{k}_2.\mathbf{r}' - i\omega_{\mathbf{k}_1}t - i\omega_{\mathbf{k}_2}t') + c.c.], \tag{2}$$

where we have written $C_{12} \equiv C_{\mathbf{k}_1\lambda_1\mathbf{k}_2\lambda_2}$ and $\hat{\epsilon}_{12} \equiv \hat{\epsilon}_{\mathbf{k}_1\lambda_1\mathbf{k}_2\lambda_2}$. It is obvious that in the energy density of this nonlocal field, contributions from the cross terms between two modes $\mathbf{k}_1\lambda_1$ and $\mathbf{k}_2\lambda_2$ vanish unless the two modes are strongly phase-correlated. Denoting the population fractions of the independent modes and of such correlated-pair modes by a_I and

[a] This feature is characterised by the appearance of the signal of doubly-charged ions in an MPI spectra *before the singly-charged ion signal has saturated.*

a_{II}, respectively, use of standard methods in QED leads to the following expression for the field energy in a volume Ω :

$$W = a_{\mathrm{I}}^2 W_{\mathrm{I}} + a_{\mathrm{II}}^2 W_{\mathrm{II}}, \tag{3}$$

where W_{I} is the usual energy summed over single modes, and

$$W_{\mathrm{II}} = \frac{1}{2}\sum_{\mathbf{k}\lambda}(Y_{\mathbf{k}\lambda}^2 + 4\omega_k^2 Z_{\mathbf{k}\lambda}^2), \tag{4}$$

where $Y_{\mathbf{k}\lambda} = -i\frac{\sqrt{\Omega/\pi}}{2c}2\omega_{\mathbf{k}}(C_{\mathbf{k}\lambda} - C_{\mathbf{k}\lambda}^*)$, $\quad Z_{\mathbf{k}\lambda} = \frac{\sqrt{\Omega/\pi}}{2c}(C_{\mathbf{k}\lambda} + C_{\mathbf{k}\lambda}^*)$. (The simplified notation $\mathbf{k}\lambda$ for the mode subscripts actually stands for a correlated mode pair $(\mathbf{k}_1\lambda_1, \mathbf{k}_2\lambda_2)$, and $2\omega_{\mathbf{k}}$ stands for $\omega_{\mathbf{k}_1} + \omega_{\mathbf{k}_2}$.) Ultimately, after canonical quantization, the nonlocal field energy is obtained in terms of a new pair of creation and annihilation opearators :

$$W_{\mathrm{II}} = \frac{1}{2}\sum_{\mathbf{k}\lambda}\hbar\omega_{\mathbf{k}}(b_{\mathbf{k}\lambda}b_{\mathbf{k}\lambda}^\dagger + b_{\mathbf{k}\lambda}^\dagger b_{\mathbf{k}\lambda}), \tag{5}$$

where $b_{\mathbf{k}\lambda}, b_{\mathbf{k}\lambda}^\dagger = \frac{1}{\sqrt{2\hbar\omega_{\mathbf{k}}}}(2\omega_{\mathbf{k}} Z_{\mathbf{k}\lambda} \pm iY_{\mathbf{k}\lambda})$ are *two-photon annihilation/creation operators*, which obey the commutation relations

$$[b_{\mathbf{k}\lambda}, b_{\mathbf{k}'\lambda'}^\dagger] = 2\delta_{\mathbf{k}\mathbf{k}'}\delta_{\lambda\lambda'}, \qquad [b_{\mathbf{k}\lambda}, b_{\mathbf{k}'\lambda'}] = [b_{\mathbf{k}\lambda}^\dagger, b_{\mathbf{k}'\lambda'}^\dagger] = 0. \tag{6}$$

The non-local field can now be written as

$$\mathcal{A}(x, x') = \sum_{\mathbf{k}\lambda}\sqrt{\frac{2\pi\hbar c^2}{\Omega\omega_{\mathbf{k}}}}\hat{\epsilon}_{\mathbf{k}\lambda}(\hat{r}, \hat{r}')[b_{\mathbf{k}\lambda} f(x, x') + b_{\mathbf{k}\lambda}^\dagger f^*(x, x')], \tag{7}$$

where $f(x, x')$ is the phase term in eqn. (2).

III. Discussion

From what has been said above, the term W_{II} in the field energy becomes important only when the modes $\mathbf{k}_1\lambda_1$ and $\mathbf{k}_2\lambda_2$ at the space-time points (x, x') are very closely matching in phase. Thus in the matrix element (1), the nonlocal effects will be prominent only under conditions of strong second-order phase coherence; otherwise the physics will be given by ordinary QED. (Mathematically it is trivially done by replacing $\mathcal{A}(x, x')$ in (1) by $\mathbf{A}(x)\delta(x - x')$ followed by the usual quantization of $\mathbf{A}(x)$.)

As has been shown in [2] and [3], this second-order coherence condition can be realized in practice only when the phase difference $\Delta\phi = \omega\delta t$ between two photon absorption/emission events by atomic electrons is vanishingly small. (Here δt is the time gap between the two photon-absorption events.) For thermal or laser light this can happen only when the photon density is orders of magnitude higher than the electron density, and this obviously sets a *frequency-dependent lower limit* to the radiant energy flux (i.e., intensity) for this coherence condition to be satisfied. For squeezed light, say from an optical parametric oscillator, absorption of a pair of *signal* and *idler* photons can meet this condition. (Recently it has been shown [7,8] that this condition can be satisfied under cosmological environments as well.) On the other hand, multiphoton processes that depend directly on electron correlation (e.g., "direct" double ionization), would occur only below a *frequency-independent upper limit* to the intensity, because the electric fields at high laser intensities approach the Coulomb fields within an atom and thus weaken the electron correlation. Obviously these two requirements set some restrictions upon the required laser parameters, e.g., intensity and wavelength, for such processes to occur. Thus, as already shown in [2] and [3], DDI would be favoured at long (within near infra-red) wavelengths and short laser pulses. This feature has been observed earlier as well as recently [9].

Matrix elements for such correlated two-photon absorption by one-electron as well as two-electron atoms have been worked out in [3], and many physical consequences of this correlated two-photon absorption by atoms in radiation fields have been explicitly described/predicted in [2], [3], [7] and [8]. It has been demonstrated in [2] and [3] that a crude estimate for the abovementioned frequency-independent upper limit of the laser intensity is of the order[b] $\sim 10^{13}$ W/cm^2, upto which electron correlation in a multi-electron atom would be maintained; above this (admittedly crude) intensity limit, the laser electric field tends to approach the Coulomb fields within the atom, whereby the electrons find it difficult to maintain their mutual correlation.

Consider now the ATI process. Each (outer) electron absorbs more photons than are necessary for photoionization, and thus the photoelectron spectra shows peaks separated by an energy gap= $\hbar\omega$ in addition to the primary peak. Note, however, that electron correlation effects can show up in these spectra if and only if two electrons can move *concertedly* during the intervening period between the two individual events of photon absorption by the two electrons. This is only possible if the electric field strength of the laser light is within the above intensity limit. The intensities used in [1] do not meet this criterion, as shown already; hence correlation effects cannot be expected in the ATI spectra of [1].

On the other hand, note that in a multiphoton absorption process, the atomic electron/s during the two-photon excitation is/are acting as a device that measures the phase of the incident radiation to within an accuracy $\Delta\phi = \omega\delta t$. (As mentioned above, δt is the time gap between the absorption of the two photons by the electron/s, and obviously decreases with increasing photon flux, i.e. intensity.) Because of the photon number-phase uncertainty relation $\Delta N \Delta\phi \geq 1$, the number of the absorbed photons would be uncertain by an amount $\delta N \sim 1/(\omega\delta t)$, and this would shift the photoelectron spectra along the energy axis through a distance $\Delta N\hbar\omega$, resulting in the damping/suppression of ΔN lower-order peaks. With increase in intensity, the phase gap $\Delta\phi$ decreases, leading to damping and ultimately suppression of more and more lower-order ATI peaks. The important point to note is that

[b] This number depends on the atom in question; thus, for Argon, it is aboout 22 times less, as shown already in § I.

this phenomenon of damping and/or suppression of peaks *does not depend on the electron correlation* because, in case two or more electrons are present, *they can act individually as such phase-measuring devices*, and hence δt can refer either to the time gap between two successive photon absorptions by the same electron, or to the time gap between the photon absorptions by two different electrons, or both. From the intensity limit given above, it is seen that at the intensities used in the experiments in [1], the electron correlation is getting progressively weakened; however, because the electrons can act individually as "devices" for measuring the phase, this weakening of the correlation does not affect the structure (i.e., peaks) in the photoelectron spectra. The spectra in [1] show no particular damping or quenching of the lower-order peaks, but in the website spectra [10] which show the complete spectrum over the 0-40 eV range, the lower-order regions at the highest intensities do appear flatter than those at the low intensities. Even if peak quenching does occur, it would not be affected by electron correlation, as has been explicitly shown in [2] and [3].

In this work we have considered correlated two-photon absorption events in the interaction of radiation with atoms. Further work on the extension of this approach to the correlated 2n-photon absorption/emission processes in atoms/molecules is in progress.

References

[1] M J Nandor, M A Walker, L D van Woerkom and H G Muller, Phys. Rev. A **60**, R1771 (1999).

[2] (a) T K Rai Dastidar and K Rai Dastidar in *Book of Abstracts, XI Int. Conf. At. Phys.* (Paris 1988), p. I-23 ;
(b) K Rai Dastidar in *Adv. At. Mol. Phys.*, ed. M S Z Chaghtai (Today's and Tomorrow's Publishers, New Delhi, 1992), p. 49.

[3] T K Rai Dastidar and K Rai Dastidar, Mod. Phys. Lett. A **13**, 1265 (Errata **13**, 2247) (1998).

[4] G Wannier, Phys. Rev. **90**, 817 (1953).

[5] N P Georgiades, E S Polzik, K Edamatsu, H J Kimble and A S Parkins, Phys. Rev. Lett. **75**, 3246(1995); Z Ficek and P D Drummond Phys. Today, September, p-34 (1997).

[6] I M Gel'fand and G E Shilov, *Generalised Functions* (Academic Press, 1964) Vol. 1, Chap. 2, Eq. (1.3)(1).

[7] T K Rai Dastidar, Mod. Phys. Lett. A **14**, 1193 (1999).

[8] T K Rai Dastidar, Mod. Phys. Lett. A **14**, 2499 (1999)..

[9] M Propopapas, C H Keittel and P L Knight, Rep. Prog. Phys. **60**, 389 (1997).

[10] www.physics/ohio-state.edu/~lvw/what/singlee/singlee.html (see ref. 31 in [1]).

Theoretical Physics 2002, Part 1
Edited by Thomas F. George and Henk F. Arnoldus, pp. 27-56
ISBN 1-59033-435-3

Jaynes-Cummings Models Driven by Classical Fields

A. Ya. Kazakov*

Abstract

An atom interacting simultaneously with a quasiresonant classical field and a quantized mode, which is in one-photon or two-photon resonance with the same atomic transition, is under consideration. The same exchange by photons between the different components of the field takes place in this situation. There is some natural hierarchy of parameters - the classical components of the external field is usually much stronger then the quantized one. This is the basis for the appropriate averaging procedure, which leads to the description of the solution of the initial value problem in fully analytic terms. The process of generation of the quantized field, the squeezing of the generated radiation and cooperative effects (when N identical atoms interact with quasiresonant classical and quantized modes) are under investigation.

1 Introduction

The Jaynes-Cummings models (JCM) that describe a single-mode quantized field's interaction with a two-(three-)level atom play a very significant role both in theoretical and experimental investigations of cavity quantum electrodynamics [1-3]. These models exhibit some interesting quantum phenomena such as the revival of atomic population, the generation of squeezed quantized radiation and so on. A considerable number of works on various aspects of these models can be found in the literature. Some papers also deal with the two-photon JCM (see, for instance, [4], [5] and references therein).

Recently, the various aspects of the one-photon and two-photon JCM driven by external classical fields have been discussed [8]-[14]. The external classical fields can be considered as useful tools to govern the dynamics of the quantum system. Their characteristics, such as frequencies and amplitudes of the separate classical harmonics, can be easily controlled, and this fact gives us the possibility to determine the main features of the dynamics of the system under consideration.

The main goal of this paper is to study the dynamics of the JCM driven by classical fields. Generally speaking, the classical fields and the quantized ones differ

*Department of Higher Mathematics, State University of AeroSpace Instrumentation, 190000, Bol'shaya Morskaya, 67, S.-Petersburg, Russia. E-mail: a_kazak@mail.ru

greatly by the number of photons, and if the corresponding coupling constants are of the same order of magnitude, then the corresponding Rabi parameters for the classical and quantized fields differ in the same way. This is the key fact for our further considerations. Taking this into account, we can separate the "slow" dynamics of our system after some averaging procedure and solve the corresponding initial value problems analytically. This averaging procedure is similar to the procedure of multiple scales [15], its analogue was called in [8] the second rotating-wave approximation. Note, that our asymptotic problem differs from the problem discussed in [16]: the physical origin of the large parameter and the physical situation as a whole are fairly different in our situation as compared to [16]. After averaging, we treat the slow dynamics with the help of the Fock-Bargmann representation. This framework reduces our problems to partial differential equations, for which we seek a general solution. So, the additional aim of this paper is to present the analytical formalism convenient for the investigation of the quantum system interacting with the strong quasiresonant classical fields.

Some variants of this model (when the classical field is monochromatic under some restrictions on the parameters of the system) were recently partly discussed. The situation when the classical component is in strict resonance with the atomic transition was considered in [8]. In papers [9], [10] this system was examined under the assumption that both components (the classical and the quantized ones) are in strict resonance with the atomic transition. In these works the driven JCM was studied in another physical sense, namely, the quantized mode was considered as a probe field for the atomic state. Moreover, in this paper we discuss another resonance situation (see details below).

This paper is organized as follows. In the next section we present the basic formalism. We describe there the averaging procedure which separates the fast oscillations due to the interaction of an atom with the classical field from the slow evolution which corresponds to the interaction of "an atom dressed by a classical field" with the quantized mode. This procedure (for the polychromatic and monochromatic classical fields) produces the averaged Hamiltonian which is the basis of the further considerations. In section 3 we discuss the one-photon JCM interacting with a classical monochromatic field and we describe the analytical solution of the initial value problem. With the help of the Fock-Bargmann representation we transform the corresponding Schrödinger equations into the partial differential equations. Using suitable analytic tools, we describe the solutions of these problems in explicit terms. In section 4 we examine the situation when N identical atoms interact simultaneously with the monochromatic classical mode and the quantized mode. We show that in this situation there is a certain cooperative phenomenon which is similar to the well-known superradiance. In section 5 we consider the dynamics of the two-photon JCM driven by a monochromatic classical field. We discuss the quantum-statistical characteristics of the radiation generated in the quantized cavity and the squeezing of this radiation. The paper concludes with a brief summary.

The details of the averaging procedure and some calculations are discussed in the Appendix.

2 Basic Formalism

2.1 Main notations and averaging procedure

We start with the Hamiltonian in the dipole and rotating-wave approximations:

$$\mathbf{H} = \omega \mathbf{a}^{+}\mathbf{a} + \kappa \mathbf{J}_0 + \zeta \left\{ (\mathbf{a}^{+})^{K} \mathbf{J}_{-} + \mathbf{a}^{K}\mathbf{J}_{+} \right\} + \tag{1}$$

$$+\mu \left[\mathbf{J}_{-} g(t) \exp(i\Omega t) + \mathbf{J}_{+}\overline{g(t)} \exp(-i\Omega t) \right]$$

where $\mathbf{a}^{+}$ and $\mathbf{a}$ are the creation and annihilation operators for the cavity mode, Ω is a carrier (optical) frequency of the external classical field, function $g(t)$ describes its modulation, μ is the corresponding Rabi parameter (this means that $\mid g(t) \mid \leq 1$ and there are such t_k, $k = 1, 2, 3...$, that $\mid g(t_k) \mid = 1$), the operators

$$\mathbf{J}_0 = diag\{1, -1\}, \mathbf{J}_{-} = \mathbf{J}_{+}^{T} = \begin{pmatrix} 0 & 0 \\ 1 & 0 \end{pmatrix},$$

describe the two-level atom interacting with the external quasiresonant field and satisfy the commutation relations,

$$[\mathbf{a}, \mathbf{a}^{+}] = 1, [\mathbf{J}_0, \mathbf{J}_{-}] = -2\mathbf{J}_{-}, [\mathbf{J}_0, \mathbf{J}_{+}] = 2\mathbf{J}_{+}.$$

The parameter ζ characterizes the intensity of the atom's interaction with the cavity mode and $(...)^{*}$ means the operator conjugation. It follows from the standard commutation relations that for any analytical function $q(s)$

$$\mathbf{a}q(\mathbf{a}^{+}\mathbf{a}) = q(\mathbf{a}^{+}\mathbf{a} + 1)\mathbf{a}, \mathbf{a}^{+}q(\mathbf{a}^{+}\mathbf{a}) = q(\mathbf{a}^{+}\mathbf{a} - 1)\mathbf{a}^{+}, \tag{2}$$

$$\mathbf{J}_{-}q(\mathbf{J}_0) = q(\mathbf{J}_0 + 2)\mathbf{J}_{-}, \mathbf{J}_{+}q(\mathbf{J}_0) = q(\mathbf{J}_0 - 2)\mathbf{J}_{+}.$$

Below we discuss the situations $K = 1$, i.e., the one-photon interaction, and $K = 2$, i.e, the two-photon interaction of the quantized mode with an atom. The Hamiltonian (1) is written using the rotating-wave approximation (RWA). It means that the optical frequency greatly exceeds other frequency parameters of the problem. Similar Hamiltonians (for the $g(t) \equiv 1$, $K = 1$, under conditions like $\omega = \Omega$ or $\omega = 2\kappa = \Omega$) were used in [8] and [14].

We are interested in the dynamics of the wave function, which is governed by the Schrödinger equation:

$$i\frac{\partial}{\partial t}\Psi = \mathbf{H}\Psi. \tag{3}$$

The Hamiltonian (1) contains several temporal parameters determined by the characteristics of the atom and the fields. They are R, the Rabi parameter of the classical field, $R \sim \mu$, the Rabi parameter of the quantized mode $R_q \sim \zeta\sqrt{n(t)}$, where $n(t)$ is the population of the quantized radiation (the average number of photons in the quantized mode), and δ, the characteristic frequency of the modulation $g(t)$. As it was stated above, we assume, that

$$R, \delta >> R_q. \tag{4}$$

Note, that $n(t)$ is not known a priori and we have to check for the validness of (4) a posteriori.

We call fast (slow) the functions which essentially change over temporal intervals like R^{-1} (R_q^{-1}). Our nearest goal is to separate the fast oscillations caused by the interactions of atom with classical field. In other words we are going to derive the Hamiltonian which is "averaged" with respect to the fast oscillations. As a first step we use the known transformation:

$$\Psi(t) = \exp\left[-i\omega t(\mathbf{a}^{+}\mathbf{a} + K\mathbf{J}_0/2)\right]\Phi(t). \tag{5}$$

It separates the optical frequency and we get:

$$i\frac{\partial}{\partial t}\Phi = \Big\{(\kappa - K\omega/2)\mathbf{J}_0 + \zeta\left[(\mathbf{a}^{+})^{K}\mathbf{J}_{-} + \mathbf{a}^{K}\mathbf{J}_{+}\right] + \tag{6}$$

$$+\mu\left[\mathbf{J}_{-}g(t)\exp\left[-i(K\omega-\Omega)t\right] + \mathbf{J}_{+}\overline{g(t)}\exp\left[i(K\omega-\Omega)t\right]\right]\Big\}\Phi.$$

Further, let 2×2−matrix $\Xi(t)$ be the solution of some auxiliary initial value problem:

$$i\frac{\partial}{\partial t}\Xi = \Big\{(\kappa - K\omega/2)\mathbf{J}_0 + \mu\left[\mathbf{J}_{-}g(t)\exp\left[-i(K\omega-\Omega)t\right] + \mathbf{J}_{+}\overline{g(t)}\exp\left[i(K\omega-\Omega)t\right]\right]\Big\}\Xi, \tag{7}$$

$$\Xi(0) = I, \tag{8}$$

where I is the 2×2 unity matrix. In the general case we can not construct the solution of (7) and (8) in explicit analytic terms.

We find a solution of (6) in the form

$$\Phi(t) = \Xi(t)\varphi(t). \tag{9}$$

As will be shown further, the relation (9) describes the wave function as a product of the fast and slow factors. Using the relation (7), we get an equation for function $\varphi(t)$:

$$i\frac{\partial}{\partial t}\varphi(t) = \zeta\Xi(t)^{-1}\left[(\mathbf{a}^{+})^{K}\mathbf{J}_{-} + \mathbf{a}^{K}\mathbf{J}_{+}\right]\Xi(t)\varphi(t). \tag{10}$$

Thus, we pass from the wave function $\Psi(t)$ to $\varphi(t)$ using the substitution:

$$\Psi(t) = \exp\left[-i\omega t(a^{+}a + K\mathbf{J}_0/2)\right]\Xi(t)\varphi(t),$$

and, as follows from the condition (8),

$$\varphi(0) = \Psi(0).$$

Equation (10) is "simpler" than equation (3), because its right-hand side is proportional to the (relatively) small parameter R_q . This fact gives us the possibility to derive the suitable asymptotic procedure to construct its solution.

If we introduce the time scaling $\tau = t(R_q)^{-1}$, then equation (10) will be similar to equation (51) from Appendix I. But we will use the initial notation, keeping in mind that we can use a procedure analogous to that presented in Appendix I. From the formal point of view it means that in order to separate the "slow" quantum dynamics, it is necessary to average relation (10) with respect to the fast oscillations which are connected with the interaction of the atom with the classical field. Thus, we find the leading asymptotic term of our solution as a solution of the equation:

$$i\frac{\partial}{\partial t}\varphi(t) = \zeta\left\langle\Xi(t)^{-1}\left\{(\mathbf{a}^{+})^{K}\mathbf{J}_{-} + \mathbf{a}^{K}\mathbf{J}_{+}\right\}\Xi(t)\right\rangle\varphi(t). \tag{11}$$

Here $\langle ... \rangle$ means that the "fast" harmonics whose frequencies are of the order of R are omitted.

We can rewrite (11) in a more explicit form using the special structure of (7). Namely, let

$$\psi(t) = \begin{pmatrix} \psi_2(t) \\ \psi_1(t) \end{pmatrix},$$

which will be the solution of the initial value problem:

$$i\frac{\partial}{\partial t}\psi = \left\{(\kappa - K\omega/2)\mathbf{J}_0 + \mu\left[\mathbf{J}_{-}g(t)\exp\left[-i(K\omega - \Omega)t\right] + \mathbf{J}_{+}\overline{g(t)}\exp\left[i(K\omega - \Omega)t\right]\right]\right\}\psi, \tag{12}$$

$$\psi(0) = \begin{pmatrix} 1 \\ 0 \end{pmatrix}.$$

Then the solution of initial value problem (7), (8) is

$$\Xi(t) = \begin{pmatrix} \psi_2(t) & -\overline{\psi_1(t)} \\ \psi_1(t) & \overline{\psi_2(t)} \end{pmatrix},$$

and

$$\Xi^{-1}(t) = \begin{pmatrix} \overline{\psi_2(t)} & \overline{\psi_1(t)} \\ -\psi_1(t) & \psi_2(t) \end{pmatrix}.$$

In this case equation (11) reads:

$$i\frac{\partial}{\partial t}\varphi(t) = \tag{13}$$

$$= \zeta \left\langle \begin{pmatrix} \left(\mathbf{a}^+\right)^K \overline{\psi_1(t)}\psi_2(t) + \mathbf{a}^K \psi_1(t)\overline{\psi_2(t)} & \mathbf{a}^K \overline{\psi_2^2(t)} - \left(\mathbf{a}^+\right)^K \overline{\psi_1^2(t)} \\ \left(\mathbf{a}^+\right)^K \psi_2^2(t) - \mathbf{a}^K \psi_1^2(t) & -\left(\mathbf{a}^+\right)^K \overline{\psi_1(t)}\psi_2(t) - \mathbf{a}^K \psi_1(t)\overline{\psi_2(t)} \end{pmatrix} \right\rangle \varphi(t).$$

As was mentioned above, this system contains three frequency scales: the first one is determined by the Rabi parameter of the classical field R, the second one is determined by the width of the spectrum of modulation of the external field, Λ, and the third one is connected with R_q. The relation between R and R_q was introduced above, but we did not fix the relation between R and Λ. The result of averaging in (13) depends on the last relation too. We will fix the corresponding relation further.

Note, that the operators $\Xi(t)$ and $\exp\left[-i\omega t(\mathbf{a}^+\mathbf{a} + K\mathbf{J}_0/2)\right]$ are unitary matrices. The Fock operators $\mathbf{a}$ and $\mathbf{a}^+$ commute with the matrix operators $\exp(-i\omega t\mathbf{J}_0/2)$, $\Xi(t)$. This means that if G is a Fock operator polynomial with respect to $\mathbf{a}^+\mathbf{a}$ and we calculate $< G >$, then

$$< G > = < \overline{\Psi(t)}, G\Psi(t) > = < \overline{\varphi(t)}, G\varphi(t) > . \tag{14}$$

2.2 Polychromatic external field

Equation (13) can be written in a more explicit form in the case of a polychromatic external field, when function $g(t)$ is periodic (this means that the harmonics of the classical external field are equidistant); we denote the frequency of modulation of the external field as D. We suppose here and below, that D has the same order of magnitude as R. In this case we can use the Floquet solutions of (12). Let the vector function $\psi(t)$ be such a solution. This means that

$$\psi(t) = diag\left\{\exp\left[i(K\omega - \Omega)t/2\right], \exp\left[-i(K\omega - \Omega)t/2\right]\right\}\exp(i\nu t)\begin{pmatrix} P_2(t) \\ P_1(t) \end{pmatrix}. \tag{15}$$

Here, $P_2(t)$, $P_1(t)$ are periodic functions ($P_k(t + 2\pi/D) = P_k(t)$, $k = 1, 2$), and ν is a Floquet parameter. Substituting relation (15) in equation (13) gives:

$$\psi_1(t)\overline{\psi_2(t)} = \exp\left[-i(K\omega - \Omega)t\right] P_1(t)\overline{P_2(t)},$$

$$\psi_2^2(t) = \exp\left[i(K\omega - \Omega)t + 2i\nu t\right] P_2^2(t),$$

$$\psi_1^2(t) = \exp\left[-i(K\omega - \Omega)t + 2i\nu t\right] P_1^2(t).$$

The different elements of the matrix coefficient in the right-hand side of relation (13) can be (or can not be) equal to zero. If a certain element contains the fast harmonics only, then the averaging procedure eliminates it.

It follows from our assumptions that $< \psi_1(t)\overline{\psi_2(t)} >= 0$, if

$$|(K\omega - \Omega) - K_1 D| >> \zeta, K_1 = 0, \pm 1, \pm 2... \qquad (16)$$

Further, if

$$|(K\omega - \Omega) + 2\nu - K_2 D| >> \zeta, K_2 = 0, \pm 1, \pm 2..., \qquad (17)$$

then $< \psi_2^2(t) >= 0$; if

$$|(K\omega - \Omega) - 2\nu - K_3 D| >> \zeta, K_3 = 0, \pm 1, \pm 2..., \qquad (18)$$

then $< \psi_2^2(t) >= 0$. Considering the other coefficients of system (13), one can conclude that there are three possibilities.

1. Relations (18) and (16) are valid, and relation (17) is invalid. Then the off-diagonal coefficients are not equal to zero, and we get:

$$i\frac{\partial}{\partial t}\varphi(t) = \zeta U \begin{pmatrix} \mathbf{0} & \mathbf{a}^K \overline{\kappa} \\ (\mathbf{a}^+)^K \kappa & 0 \end{pmatrix} U^{-1}\varphi(t), \qquad (19)$$

$$\kappa = < \exp\left[i(K\omega - \Omega)t + 2i\nu t\right] P_2^2(t) >,$$

$$U = \Xi(0) = \begin{pmatrix} \psi_2(0) & -\overline{\psi_1(0)} \\ \psi_1(0) & \overline{\psi_2(0)} \end{pmatrix}.$$

Note that in this case $\Xi(0) \neq I$ and we have to take into account the corresponding modification of our relations. This Hamiltonian differs from the usual (maybe, many-photon) JCM Hamiltonian (after transformation (5)) by a trivial rotation only. Note that in the physical system, governed by the JCM Hamiltonian, the number of excitations is conserved and there is some "deformation" of the cavity state as a result of the interaction with the atom. In order to describe the generation of photons in the cavity in this situation, it is necessary to consider the next asymptotic term. But we will discuss here the leading asymptotic term only.

2. Relation (18) is violated. Then we get the usual (maybe, many-photon) Jaynes-Cummings system too.

3. Relations (17) and (18) are valid, and relation (16) is invalid. Then only the diagonal coefficients in the right-hand side of (13) are not equal to zero, and we get the system:

$$i\frac{\partial}{\partial t}\varphi(t) = \zeta U \begin{pmatrix} \left(\mathbf{a}^{+}\right)^{K}\kappa + \mathbf{a}^{K}\overline{\kappa} & \mathbf{0} \\ \mathbf{0} & -\left(\mathbf{a}^{+}\right)^{K}\kappa - \mathbf{a}^{K}\overline{\kappa} \end{pmatrix} U^{-1}\varphi(t), \qquad (20)$$

$$\kappa = < \exp\left[-i(K\omega - \Omega)t\right] P_1(t)\overline{P_2(t)} > .$$

Namely this situation we consider below.

Further possibilities arise when a pair from the relations (16), (17), and (18) are violated. This can be so in the case of parametric resonances (see [19-21]). But we will discuss these situations elsewhere.

2.3 The monochromatic external field

If $g(t) \equiv 1$, then we can connect the parameters of the averaged Hamiltonian with the parameters of the external field in a more explicit form. Namely, in this case

$$\Xi(t) = \exp\left[i(K\omega - \Omega)t\mathbf{J}_0/2\right]\mathbf{U}\exp\left[-iRt\mathbf{J}_0\right]\mathbf{U}^{-1},$$

$$\mathbf{U} = \begin{pmatrix} \mu & \Delta - R \\ R - \Delta & \mu \end{pmatrix}, R = \sqrt{\mu^2 + \Delta^2}, \Delta = \kappa - \Omega/2.$$

Simple calculations lead to the conclusion that the operator $\Xi(t)^{-1}\left[\left(\mathbf{a}^{+}\right)^{K}\mathbf{J}_{-} + \mathbf{a}^{K}\mathbf{J}_{+}\right]\Xi(t)$ contains the following harmonics: $\pm(K\omega - \Omega)$, $\pm(K\omega - \Omega \pm 2R)$. So, as in the previous item, we have three situations when the averaging leads to the nontrivial effects.

1. $|R - (K\omega - \Omega)/2| = |\sigma| << R$. In this case the averaged Hamiltonian has the form:

$$\mathbf{H}_1 = \zeta\mu^2 D^{-1}\mathbf{U}\begin{pmatrix} 0 & \mathbf{a}^{K}\exp\left[2i\sigma t\right] \\ \left(\mathbf{a}^{+}\right)^{K}\exp\left[-2i\sigma t\right] & 0 \end{pmatrix}\mathbf{U}^{-1},$$

$$D = \det\mathbf{U} = \mu^2 + (R - \Delta)^2.$$

2. $|(K\omega - \Omega)/2 + R| = |\eta| << R$. In this case the averaged Hamiltonian in the right-hand side of (11) has the form:

$$\mathbf{H}_2 = -\zeta(R-\Delta)^2 D^{-1}\mathbf{U}\begin{pmatrix} 0 & \left(\mathbf{a}^+\right)^K \exp\left[2i\eta t\right] \\ \mathbf{a}^K \exp\left[-2i\eta t\right] & 0 \end{pmatrix}\mathbf{U}^{-1}$$

As before, these situations are not interesting for us.

3. $|K\omega - \Omega|/2 = |\nu| << R$. The averaging of the right-hand side of (11) yields:

$$\mathbf{H}_3 = \zeta\mu(R-\Delta)D^{-1}\left(\left(\mathbf{a}^+\right)^K e^{2i\nu t} + \mathbf{a}^K e^{-2i\nu t}\right)\mathbf{U}\mathbf{J}_0\mathbf{U}^{-1}.$$

Comparing this Hamiltonian with (20), we conclude, that the usage of the monochromatic field gives the same possibilities as the polychromatic field, if we do not take into account the parametric resonances.

Note that the eigenvectors of the matrix operator $\mathbf{U}\mathbf{J}_0\mathbf{U}^{-1}$ are

$$\mathbf{e}_1 = \frac{1}{\sqrt{D}}\begin{pmatrix} \mu \\ R-\Delta \end{pmatrix}, \mathbf{e}_2 = \frac{1}{\sqrt{D}}\begin{pmatrix} \Delta - R \\ \mu \end{pmatrix},$$

and $\varepsilon_k = (-1)^{k+1}, k = 1, 2$, are the corresponding eigenvalues. It follows from the last relation that in this case the dynamics of the system splits up to the elementary rotation on two identical one-dimensional problems. Thus, we can restrict ourselves to the investigation of the system governed by the averaged Hamiltonian

$$\mathbf{H}_4 = \rho\left(\left(\mathbf{a}^+\right)^K e^{2i\nu t} + \mathbf{a}^K e^{-2i\nu t}\right), \tag{21}$$

where $\rho = \zeta\mu(R-\Delta)/D$. So, our Schrödinger equation can be reduced to the following form:

$$i\frac{\partial\varphi}{\partial t} = \rho\left(\left(\mathbf{a}^+\right)^K e^{2i\nu t} + \mathbf{a}^K e^{-2i\nu t}\right)\varphi. \tag{22}$$

Remark

Resonance conditions 1 and 2 were discussed in papers [8] and [12]. Evidently, these situations differ greatly from situation 3, both mathematically and physically.

3 One-Photon Interaction

First we discuss the one-photon interaction, when $K = 1$. In this case equation (22) reads:

$$i\frac{\partial\varphi}{\partial t} = \rho\left(\mathbf{a}^+ e^{2i\nu t} + \mathbf{a}e^{-2i\nu t}\right)\varphi. \tag{23}$$

As will be shown below, the Fock-Bargmann representation is a very handy formalism for the solving of the corresponding initial value problems. We have to use the substitutions $\mathbf{a}^+ \longmapsto z$, $\mathbf{a} \longmapsto D_z$, where D_z means the differentiation with respect to z. According to [17], for any Fock operator G its value can be calculated with help of the relation

$$< G >=< \overline{\varphi}, G\varphi >= \int dz d\overline{z} \exp[-z\overline{z}] \overline{\varphi(z,t)} G\varphi(z,t). \tag{24}$$

Equation (23) transforms in this framework into

$$i\frac{\partial \varphi(z,t)}{\partial t} = \rho \left[z \exp(2i\nu t) + D_z \exp(-2i\nu t) \right] \varphi(z,t). \tag{25}$$

It is a partial differential equation of the first order. Its solution can be constructed using the substitutions $z\exp(2i\nu t) = y, t = \tau$. Omitting some simple but cumbersome details, we write the final result only:

$$\varphi(z,t) = \exp\left\{ \frac{z\rho}{2\nu} [1 - \exp(2i\nu t)] + \right.$$

$$\left. + \frac{\rho^2}{4\nu^2} [\exp(-2i\nu t) + 2i\nu t - 1] \right\} Q\left[z + \frac{\rho}{2\nu} (\exp(-2i\nu t) - 1) \right]. \tag{26}$$

Relation (26) describes the general solution of the initial value problem for equation (25). The function $Q(z)$ can be determined from the initial conditions. If $\Psi(z,0) = \Phi_0(z)\mathbf{e}_k$, where $\mathbf{e}_k$ is the eigenvector of the matrix $\mathbf{U}\mathbf{J}_0\mathbf{U}^{-1}$, then $Q(z) = \varphi(z,0) = \Phi_0(z)$. At $\nu = 0$ expression (26) gives

$$\varphi(z,t) = \exp\left[-i\rho z t - \rho^2 t^2/2 \right] Q(z - i\rho t).$$

Let us discuss the initial state situation when the quantized mode contains m photons. As it follows from our results, in this case $Q(z) = z^m/\sqrt{m!}$. Then

$$\varphi(z,t) = \exp\left\{ \frac{z\rho}{2\nu} (1 - \exp(2i\nu t)) + \right.$$

$$\left. + \frac{\rho^2}{4\nu^2} (\exp(-2i\nu t) + 2i\nu t - 1) \right\} \left[z + \frac{\rho}{2\nu} (\exp(-2i\nu t) - 1) \right]^m / \sqrt{m!}. \tag{27}$$

At first we calculate the dynamics of the population of the quantized mode, $< n(t) >=< zD_z >$. In accordance with our results we get

$$< n(t) >= \int dz d\overline{z} \exp\{-z\overline{z}\} \overline{\varphi(z,t)} z D_z \varphi(z,t). \tag{28}$$

We use below the relation which can be deduced by integration by parts:

$$\int dz d\overline{z} \exp\{-z\overline{z}\} \overline{\varphi(z,t)} z^k D_z^k \varphi(z,t) =$$

$$= \int dzd\bar{z} \exp\{-z\bar{z}\} \overline{D_z^k \varphi(z,t)} D_z^k \varphi(z,t). \tag{29}$$

and substitution

$$z = (u + i\rho \frac{\sin(\nu t)}{\nu}) \exp(-i\nu t).$$

The integral in (28) can be evaluated in explicit terms and we have:

$$< n(t) >= m + \frac{\rho^2 \sin^2(\nu t)}{\nu^2}. \tag{30}$$

Using analogous considerations we obtain:

$$< (\mathbf{a}^+)^2 \mathbf{a}^2 >= m(m-1) + 4m \frac{\rho^2 \sin^2(\nu t)}{\nu^2} + \frac{\rho^4 \sin^4(\nu t)}{\nu^4}.$$

Then we can calculate the Mandel's factor [18]:

$$M \equiv (< (\mathbf{a}^+)^2 \mathbf{a}^2 > - < \mathbf{a}^+\mathbf{a} >^2)(< \mathbf{a}^+\mathbf{a} >)^{-1} =$$

$$= m(2\frac{\rho^2 \sin^2(\nu t)}{\nu^2} - 1)(m + \frac{\rho^2 \sin^2(\nu t)}{\nu^2})^{-1}. \tag{31}$$

Let us calculate the correlation function of the second order. It is usually determined in the Heisenberg formalism as [18]:

$$g^{(2)}(t, t+\tau) \equiv \frac{< \mathbf{a}^+(t)\mathbf{a}^+(t+\tau)\mathbf{a}(t+\tau)\mathbf{a}(t) >}{< \mathbf{a}^+(t)\mathbf{a}(t) >< \mathbf{a}^+(t+\tau)\mathbf{a}(t+\tau) >}. \tag{32}$$

Since we use the Schrödinger formalism we have to transform the last expression. Let us introduce the unitary operator $U(t_2, t_1)$ which describes the evolution of the quantum system between t_1 and t_2. In these terms the correlation function is:

$$g^{(2)}(t, t+\tau) = \frac{(\overline{\varphi_0}, U^{-1}(t,0)\mathbf{a}^+ U^{-1}(t+\tau,t)\mathbf{a}^+\mathbf{a} U(t+\tau,t)\mathbf{a} U(t,0)\varphi_0)}{< n(t) >< n(t+\tau) >}. \tag{33}$$

Using the relation $\exp[\alpha z D_z]\xi(z) = \xi(z\exp(\alpha))$ which is valid for any function $\xi(z)$, we get:

$$g^{(2)}(t, t+\tau) = v(t,\tau)(m + \frac{\rho^2 \sin^2(\nu t)}{\nu^2})^{-1}(m + \frac{\rho^2 \sin^2[\nu(t+\tau)]}{\nu^2})^{-1},$$

$$v(t,\tau) = m(m-1) + \frac{\rho^4 \sin^2(\nu t) \sin^2[\nu(t+\tau)]}{\nu^4} +$$

$$+\frac{\rho^2 m}{\nu^2}\left\{\sin^2\left[\nu(t+\tau)\right] + 2\sin(\nu t)\cos(\nu t)\sin[\nu(t+\tau)] + \sin^2(\nu t)\right\}.$$

The important characteristics of the quantum-statistical properties of the quantized radiation is the function

$$q(t) \equiv \frac{\partial}{\partial \tau} g^{(2)}(t, t+\tau)\mid_{\tau=0}$$

Finally we obtain:

$$q(t) = m\rho^2 \sin(2\nu t)\left[m + 1 - \frac{\rho^2 \sin^2 \nu t}{\nu^2}\right]\nu^{-1}\left[m + \frac{\rho^2 \sin^2 \nu t}{\nu^2}\right]^{-3}$$

At $\nu = 0$ these relations read:

$$< n(t) >= m + \rho^2 t^2,$$

$$M = \frac{m(2\rho^2 t^2 - 1)}{m + \rho^2 t^2},$$

$$q(t) = 2\rho^2 m t(m + 1 - \rho^2 t^2)(m + \rho^2 t^2)^{-3}.$$

Let us discuss these results. From (30) it follows that if $\nu \neq 0$, then the population of the quantized mode oscillates and, specifically, has a maximum equal to $m + \rho^2 \nu^{-2}$. If, for instance, the initial state of the cavity mode is the vacuum state ($m = 0$), then this population reduces periodically to zero. The other characteristics of the quantized mode also have an oscillating behavior. The frequency of these oscillations ν does not depend on the initial population of the quantized mode and is determined by the difference of frequencies of the classical and quantized components of the field.

As it follows from relation (31), for $m = 0$ the radiation in the quantized mode is Poissonian. When $m\nu \neq 0$ this radiation is sub-Poissonian during one part of the period of oscillation , and is super-Poissonian on the remaining part. (If $\sqrt{2}\rho < \nu$, the radiation will be always sub-Poissonian.) Further, during the whole period there is either bunching or anti-bunching of the photons in the cavity mode (in accordance with the sign of the function $q(t)$; here we use the terminology of [18]).

As ν decreases the frequency of the oscillations mentioned above also decreases and the mean population of the quantized mode increases. The special case $\nu = 0$ corresponds to the strict resonance between the quantized and classical components. From a formal point of view it means that the period of oscillations equals infinity and there is an infinite pumping of the quantized mode.

It is not difficult to consider the situation when the initial state is a coherent one,

$$\mid \alpha >= \exp\left[- \mid \alpha \mid^2 /2 + \alpha z\right].$$

In this case

$$\varphi(z,t) = \exp\Big\{- \mid \alpha \mid^2 /2 + \alpha\left[z + \rho(\exp(-2i\nu t) - 1)/2\nu\right] +$$

$$+\rho z\left[1 - \exp(2i\nu t)\right]/2\nu + \rho^2(\exp(-2i\nu t) + 2i\nu t - 1)/4\nu^2\Big\}. \qquad (34)$$

This function is another coherent state corresponding to the parameter $\alpha\exp(i\nu t) - i\rho\sin(\nu t)/\nu$ within the accuracy of an elementary factor and rotation $z \rightarrow z\exp(-i\nu t)$. So we can transform one coherent state into another coherent state with controlled (in some limits) parameters. Further,

$$< n(t) >=\mid \alpha \mid^2 + \rho^2\sin^2(\nu t)/\nu^2 + 2\rho \mid \alpha \mid \sin(\nu t)\sin(\nu t - \arg\alpha)/\nu.$$

Note that it is possible to have periodic vanishing of the $< n(t) >$ at $\mid \alpha \mid \nu < \rho$ and a special choice of value of $\arg\alpha$. As it follows from the special structure of the function (34) in this case the Mandel factor equals zero (the radiation is Poissonian) and the relation $g^{(2)}(t, t+\tau) = 1$ is valid. At $\nu = 0$ we obtain:

$$< n(t) >=\mid \alpha \mid^2 + \rho^2 t^2 - 2\rho \mid \alpha \mid t\sin(\arg\alpha).$$

At first the population of the quantized mode decreases (if $0 < \arg\alpha < \pi/2$) and then increases quadratically.

4 Forced Superradiance

4.1 Basic notations

In the preceding section it was shown that the quasiresonant classical and quantized modes exchange photons and the atom here is a kind of a photons transmitter. The natural question arises: Will the cooperative properties manifest themselves if N identical atoms interact simultaneously with both modes? In this section we discuss this problem. We show, that the number of photons in the quantized mode oscillates, and the amplitude of the oscillation is proportional to N^2.

So, suppose that our system contains $N > 1$ identical two-level atoms interacting simultaneously with the classical monochromatic mode and with the quantized one. For the state space we take $L = F \otimes C^2 \otimes C^2 ... \otimes C^2$, where the Fock space F describes the state of the quantized mode, and N copies of C^2 describe the states of the atoms. Then L is the linear span of vectors of type $f \mid e_{k_1}, e_{k_2}, ..., e_{k_N} >$ where e_{k_n} are, as described above, eigenvectors of the matrix $\mathbf{U}\mathbf{J}_0\mathbf{U}^{-1}$, $k_m = 1, 2$.

We define the operators $\mathbf{J}_0^{(m)}, \mathbf{J}_-^{(m)}, \mathbf{J}_+^{(m)}$ and $\Xi^{(m)}(t), 1 \leq m \leq N$, as follows: they act on the m-th component of the vector $f \mid v_1 v_2, ..., v_N >$ as the operators $\mathbf{J}_0, \mathbf{J}_-, \mathbf{J}_+$ and $\Xi(t)$ respectively and they do not alter the other components. Then the Hamiltonian of our system (in the rotating-wave approximation) can be written as

$$\mathbf{H}_N = \omega \mathbf{a}^+ \mathbf{a} + \kappa \sum_{m=1}^{N} \mathbf{J}_0^{(m)} + \zeta \sum_{m=1}^{N} \left\{ \mathbf{a}^+ \mathbf{J}_-^{(m)} + \mathbf{a} \mathbf{J}_+^{(m)} \right\} + \tag{35}$$

$$+\mu \sum_{m=1}^{N} \left[\mathbf{J}_-^{(m)} \exp(i\Omega t) + \mathbf{J}_+^{(m)} \exp(-i\Omega t) \right].$$

We are interested in the dynamics of the wave function $\Psi(t)$, which is the solution of the corresponding Schrödinger equation. We separate the optical frequency by using an analog of eq. (5):

$$\Psi(t) = \exp\left[-i\omega t \left(\mathbf{a}^+ \mathbf{a} + \sum_{m=1}^{N} \mathbf{J}_0^{(m)}/2 \right) \right] \Phi(t).$$

For the function $\Phi(t)$ we get the equation:

$$i \frac{\partial \Phi(t)}{\partial t} = \left\{ (\kappa - \omega/2) \sum_{m=1}^{N} \mathbf{J}_0^{(m)} + \zeta \sum_{m=1}^{N} \left[\mathbf{a}^+ \mathbf{J}_-^{(m)} + \mathbf{a} \mathbf{J}_+^{(m)} \right] + \right.$$

$$\left. \mu \sum_{m=1}^{N} \left[\mathbf{J}_-^{(m)} \exp\left(i(\Omega - \omega)t\right) + \mathbf{J}_+^{(m)} \exp\left(i(\omega - \Omega)t\right) \right] \right\} \Phi(t).$$

The matrices $\Xi^{(m)}(t)$, $m = 1, 2, ..., N$, commute. We introduce the matrix $\Xi_N(t) = \prod_{m=1}^{N} \Xi^{(m)}(t)$. Then $\Xi_N(0)$ is the identity matrix in $C^2 \otimes C^2 ... \otimes C^2$ and $\Xi_N(t)$ is the solution of the equation

$$i \frac{\theta}{\partial t} \Xi_N(t) = \left\{ (\kappa - \omega/2) \sum_{m=1}^{N} \mathbf{J}_0^{(m)} + \mu \sum_{m=1}^{N} \left[\mathbf{J}_-^{(m)} \exp\left(i(\Omega - \omega)t\right) + \mathbf{J}_+^{(m)} \exp\left(i(\omega - \Omega)t\right) \right] \right\} \Xi_N(t).$$

Substituting $\Phi(t) = \Xi_N(t) \varphi(t)$, we arrive at the equation for function $\varphi(t)$:

$$i \frac{\partial \varphi(t)}{\partial t} = \zeta \Xi_N^{-1}(t) \sum_{m=1}^{N} \left[\mathbf{a}^+ \mathbf{J}_-^{(m)} + \mathbf{a} \mathbf{J}_+^{(m)} \right] \Xi_N(t) \varphi(t). \tag{36}$$

Note that $\varphi(0) = \Phi(0)$. The operator

$$\exp\left[-i\omega t \left(\mathbf{a}^+ \mathbf{a} + \sum_{m=1}^{N} \mathbf{J}_0^{(m)}/2 \right) \right] \Xi_N(t)$$

is unitary and commutes with the operator $\mathbf{a}^+ \mathbf{a}$. So, for the operator $G = (\mathbf{a}^+ \mathbf{a})^m$ acting in the Fock space F we have $< G >=< \overline{\Psi(t)}, G\Psi(t) >=< \overline{\varphi(t)}, G\varphi(t) >$.

4.2 The slow dynamics

The right-hand side of equation (36) contains both fast oscillations and slow oscillations, and the last ones govern the slow dynamics of the system. Discarding the fast harmonics as before, we get the equation:

$$i\frac{\partial\varphi(t)}{\partial t} = \zeta < \Xi_N^{-1}(t)\sum_{m=1}^{N}\left[\mathbf{a}^+\mathbf{J}_-^{(m)} + \mathbf{a}\mathbf{J}_+^{(m)}\right]\Xi_N(t)\varphi(t) > .$$

As before, we examine the case $| \omega - \Omega | = 2 | \nu | << R$, where R is the corresponding Rabi parameter of the classical field. This condition imposes a restriction on the difference of the frequencies of the classical and the quantized components. The averaging procedure yields in this case a nontrivial contribution, and we find that

$$\mathbf{H}_{av} = \rho\left[\mathbf{a}^+\exp(2i\nu t) + \mathbf{a}\exp(-2i\nu t)\right]\sum_{m=1}^{N}\mathbf{U}_m\mathbf{J}_0^{(m)}\mathbf{U}_m^{-1},$$

where matrix $\mathbf{U}_m$ acts as matrix $\mathbf{U}$ on the $m-$th component and does not alter the other components. This Hamiltonian controls the slow evolution of atoms "dressed by classical fields". As before, this Hamiltonian is the product of the Fock operator $\mathbf{a}^+\exp(2i\nu t) + \mathbf{a}\exp(-2i\nu t)$ and a matrix factor. Thus, if we pass to the expansion in the basis of vectors $| \mathbf{e}_{k_1}, \mathbf{e}_{k_2}, ..., \mathbf{e}_{k_N} >$, which are the eigenvectors of the matrix $\sum_{m=1}^{N}\mathbf{U}_m\mathbf{J}_0^{(m)}\mathbf{U}_m^{-1}$, the averaged Hamiltonian splits into a set of one-dimensional Hamiltonians. Then we expand the required function as

$$\varphi(t) = \sum_{\sigma}\eta_\sigma(t) \mid \mathbf{e}_{k_1}, \mathbf{e}_{k_2}, ..., \mathbf{e}_{k_N} >,$$

where the values of the functions $\eta_\sigma(t)$ belong to the Fock space, σ stands for the set of N numbers $k_1, k_2, ..., k_N$, each of which is either 1 or 2, and the sum is over all such sets σ. For each $\eta_\sigma(t)$ we get:

$$i\frac{\partial\eta_\sigma(t)}{\partial t} = \rho S_\sigma\left[\mathbf{a}^+\exp(2i\nu t) + \mathbf{a}\exp(-2i\nu t)\right]\eta_\sigma(t), \tag{37}$$

where $S_\sigma = \sum_{m=1}^{N}(-1)^{k_m+1}$, with the sum incorporating the numbers k_m that comprise the set σ. So, here S_σ is the sum over m of the eigenvalues of the operators $\mathbf{U}_m\mathbf{J}_0^{(m)}\mathbf{U}_m^{-1}$, these eigenvalues are ± 1. Note that only the factor S_σ depending on σ is present in equation (37). The solution of the initial value problem for equation (37) with $\eta_\sigma(z,0) = Q_\sigma(z)$ is described by the evident modification of relation (27):

$$\eta_\sigma(z,t) = \exp\left\{\frac{\rho S_\sigma z}{2\nu}(1 - \exp(2i\nu t)) + \right.$$

$$\left.\frac{\rho^2 S_\sigma^2}{4\nu^2}\left[\exp(-2i\nu t) + 2i\nu t - 1\right]\right\} Q_\sigma\left(z + \frac{\rho S_\sigma}{2\nu}(\exp(-2i\nu t) - 1)\right).$$

These relations make it possible to describe analytically the solution of any initial value problem.

4.3 The cooperative behavior

Let us discuss the following initial value problem: the quantized mode contains m photons initially, and all the atoms are in the initial state $v \in C^2$, with $v = \mathbf{e}_1 \cos\chi + \mathbf{e}_2 \sin\chi$. Expanding our wave function in the basis described above we arrive at the following expression for the initial values of the corresponding coefficients:

$$\eta_\sigma(z,0) = (\cos\chi)^{N-k} (\sin\chi)^k \frac{z^m}{\sqrt{m!}},$$

where k is the number of "twos" in the set σ. For a given set σ we denote this number by $\mid \sigma \mid$. Then $\eta_\sigma(z,t) = (\cos\chi)^{N-k} (\sin\chi)^k \varphi_\sigma(z,t)$, where function $\varphi_\sigma(z,t)$ is described by (27) with substitution $\rho \to \rho S_\sigma$. Note that if $\mid \sigma \mid= k$, then $S_\sigma = N-2k$, and also note that for a given k there are $\begin{pmatrix} N \\ k \end{pmatrix}$ different sets σ with given $\mid \sigma \mid$.

In accordance with our results, for the Fock operator $G = (\mathbf{a}^+\mathbf{a})^m$ we obtain in the Fock-Bargmann representation,

$$< G >=< \overline{\varphi(z,t)}, G\varphi(z,t) >= \sum_\sigma \int dzd\overline{z} \exp\left(-z\overline{z}\right) \overline{\eta_\sigma(z,t)} G\eta_\sigma(z,t) =$$

$$\sum_{k=0}^{N} \sum_{|\sigma|=k} \int dzd\overline{z} \exp\left(-z\overline{z}\right) \overline{\eta_\sigma(z,t)} G\eta_\sigma(z,t) =$$

$$\sum_{k=0}^{N} \begin{pmatrix} N \\ k \end{pmatrix} (\cos\chi)^{2N-2k} (\sin\chi)^{2k} \int dzd\overline{z} \exp\left(-z\overline{z}\right) \overline{\varphi_\sigma(z,t)} G\varphi_\sigma(z,t) \mid_{|\sigma|=k} .$$

Here we have allowed for the fact that the coefficients $\eta_\sigma(z,t)$ with the same S_σ coincide.

Let us use these results to find the number of photons in the quantized mode. The corresponding integrals were calculated earlier and we have:

$$< n(t) >=< \mathbf{a}^+\mathbf{a} >= \sum_{k=0}^{N} \begin{pmatrix} N \\ k \end{pmatrix} (\cos\chi)^{2N-2k} (\sin\chi)^{2k} \left[m + (N-2k)^2 \rho^2 \frac{\sin^2 \nu t}{\nu^2} \right].$$

Taking into account the relation

$$\sum_{k=0}^{N} \begin{pmatrix} N \\ k \end{pmatrix} (N-2k)^s (\cos\chi)^{2N-2k} (\sin\chi)^{2k} =$$

$$(\cos\chi \sin\chi)^N \left\{ \frac{d^s}{dy^s} \left[\exp(y) + \exp(-y)\right] \right\} \mid \exp y = \tan\chi,$$

we get the final formula for the population of the quantized mode:

$$< n(t) >= m + \rho^2 \frac{\sin^2 \nu t}{\nu^2} \Big[N(N-1)\cos^2 2\chi + N\Big] . \tag{38}$$

Remember that N is the number of atoms. Thus, the population of the quantized radiation which is proportional to $< n(t) >$, is proportional to N^2 . The relation (38) manifests a cooperative behavior of the atoms in our system similar to the well-known phenomenon of superradiance (see [26]). But in our case the intensity of radiation is a periodic function of time, with the period determined by the parameter ν , the difference of the frequencies between the classical and the quantized components. In contrast to "ordinary" superradiance the temporal dynamics of our system is independent of N .

5 Two-Photon Interaction

5.1 Solution of the initial value problem

In this section we discuss the situation of the two-photon interaction of the quantized mode with the two-level atom, which simultaneously interacts with the strong classical quasiresonant monochromatic field. In other words, we suppose here that $K = 2$. It should be noted here that the dynamics of the two-photon JCM was discussed before in papers [4], [5], and the driven JCM was discussed in [14]. In this case equation (22) reads:

$$i\frac{\partial \varphi}{\partial t} = \rho \left((\mathbf{a}^+)^2 e^{2i\nu t} + \mathbf{a}^2 e^{-2i\nu t} \right) \varphi,$$

which is in the Fock-Bargmann representation:

$$i\frac{\partial \varphi}{\partial t} = \rho \left(z^2 e^{2i\nu t} \varphi + e^{-2i\nu t} \varphi_{zz} \right) . \tag{39}$$

The change of variables

$$\tau = t, y = z \exp{(i\nu t)} ,$$

leads to the equation

$$i\varphi_\tau = \rho \Big[\varphi_{yy} + y^2 \varphi\Big] + \nu y \varphi_y . \tag{40}$$

The last equation can be solved by separation of variables (see [27]). But here we use the more handy approach, which gives us the possibility to obtain a general solution of the initial problem. Using these results, we can describe the quantum-statistical characteristics of the quantized radiation completely analitically. (We place the corresponding details in the Appendix.) So,

$$\varphi(z,t) = \exp\left[\frac{\beta z^2}{2}\exp(2i\nu t) + i(\nu+\beta\rho)t\right] \tag{41}$$

$$\int ds \exp\left[iz\exp(i\nu t)\,s - \frac{\rho s^2}{2(\nu+2\beta\rho)}\right]\Pi(s\exp[i(\nu+2\beta\rho)t]),$$

where $\beta = -\left(\nu+\sqrt{\nu^2-4\rho^2}\right)(2\rho)^{-1}$, function $\Pi(x)$ should be determined from the initial data, and the corresponding equation is:

$$\exp\left[\frac{\beta z^2}{2}\right]\int ds\exp\left[izs - \frac{\rho s^2}{2(\nu+2\beta\rho)}\right]\Pi(s) = \varphi(0,z) = S(z). \tag{42}$$

If the initial state is coherent, $S(z) = \exp(\alpha z)$, then we get:

$$\exp\left[\frac{\beta z^2}{2}\right]\int ds\exp\left[izs - \frac{\rho s^2}{2(\nu+2\beta\rho)}\right]\Pi(s) = \exp(\alpha z).$$

Simple but cumbersome calculations lead in this case to the following relations:

$$\Pi(s) = \frac{1}{\beta\sqrt{\pi}}\exp\left[\frac{\alpha^2}{2\beta} - \frac{i\alpha s}{\beta} - \frac{s^2}{2\beta} + \rho\frac{s^2}{2(\nu+2\beta\rho)}\right],$$

and, correspondingly,

$$\varphi(z,t) = \sqrt{\frac{2\sqrt{\nu^2-4\rho^2}}{\sigma(t)}}\exp\left\{i\nu t/2 + \frac{2\alpha z\sqrt{\nu^2-4\rho^2}\exp(i\nu t) - \alpha^2\rho\chi(t) - \rho z^2\exp(2i\nu t)\,\chi(t)}{\sigma(t)}\right\}, \tag{43}$$

where

$$\chi(t) = \exp\left(i\sqrt{\nu^2-4\rho^2}t\right) - \exp\left(-i\sqrt{\nu^2-4\rho^2}t\right),$$

$$\sigma(t) = \left(\nu+\sqrt{\nu^2-4\rho^2}\right)\exp\left(i\sqrt{\nu^2-4\rho^2}t\right) - \left(\nu-\sqrt{\nu^2-4\rho^2}\right)\exp\left(-i\sqrt{\nu^2-4\rho^2}t\right).$$

5.2 Quantum-statistical characteristics of the quantized radiation

We consider here the case $\alpha = 0$ only, since the general situation leads to essentially more complex relations. More precisely, our initial state is $\Psi(0) = \mid 0 > \mathbf{e}_1$, where $\mid 0 >$ is the vacuum state and $\mathbf{e}_1$ is the mentioned above eigenvector. We examine here the value of Fock operators only, so the vector structure of the wave function can be neglected. We consider the number of photons $N(t) = < \mathbf{a}^+\mathbf{a} >$ (the population of the quantized mode) and values $D_\pm(t) = < (\Delta X_\pm(t))^2 >$, where

$X_+(t) = [\mathbf{a}\exp(i\omega t) + \mathbf{a}^+ \exp(-i\omega t)]/2$, $X_-(t) = [\mathbf{a}\exp(i\omega t) - \mathbf{a}^+ \exp(-i\omega t)]/2i$, are the quadrature components. In more detailed form:

$$D_+(t) = \Big[2 < \mathbf{a}^+\mathbf{a} > +1+ < \mathbf{a}^2 \exp(2i\omega t) > + < (\mathbf{a}^+)^2 \exp(-2i\omega t) > - , \tag{44}$$

$$< \left[\mathbf{a}\exp(i\omega t) + \mathbf{a}^+ \exp(-i\omega t)\right] >^2\Big]/4$$

$$D_-(t) = \Big[2 < \mathbf{a}^+\mathbf{a} > +1- < \mathbf{a}^2 \exp(2i\omega t) > - < (\mathbf{a}^+)^2 \exp(-2i\omega t) > - \tag{45}$$

$$< i\left[\mathbf{a}\exp(i\omega t) - \mathbf{a}^+ \exp(-i\omega t)\right] >^2\Big]/4$$

In order to calculate these values we can use the same considerations as in the preceding section, taking into account the commutation relations (2). So,

$$N(t) = \int dz d\bar{z} \exp(-z\bar{z})\overline{\varphi(z,t)} z \frac{d}{dz}\varphi(z,t) = \int dz d\bar{z} (z\bar{z} - 1) \exp(-z\bar{z})\overline{\varphi(z,t)}\varphi(z,t),$$

$$D_\pm(t) = \frac{1}{4}\int dz d\bar{z} \left[2z\bar{z} - 1 \pm (z^2 + \bar{z}^2)\right] \exp(-z\bar{z})\overline{\varphi(z,t)}\varphi(z,t).$$

Taking into account relation (43), we can reduce these integrals to a combination of the Gaussians and obtain the following: if $|\nu| > 2\rho$, then

$$N(t) = \frac{4\rho^2}{\nu^2 - 4\rho^2}\sin^2\left(\sqrt{\nu^2 - 4\rho^2}t\right), \tag{46}$$

$$D_\pm(t) = \frac{1}{4} \pm \frac{\rho}{4(\nu^2 - 4\rho^2)}\left\{\left(\nu + \sqrt{\nu^2 - 4\rho^2}\right)\cos\left[2\left(\nu + \sqrt{\nu^2 - 4\rho^2}\right)t\right] + \right.$$

$$\left. + \left(\nu - \sqrt{\nu^2 - 4\rho^2}\right)\cos\left[2\left(\nu - \sqrt{\nu^2 - 4\rho^2}\right)t\right] - 2\nu\cos(2\nu t)\right\} + \frac{2\rho^2}{\nu^2 - 4\rho^2}\sin^2\left(\sqrt{\nu^2 - 4\rho^2 t}\right). \tag{47}$$

If $|\nu| < 2\rho$, then

$$N(t) = \frac{4\rho^2}{4\rho^2 - \nu^2}\left\{\sinh\left(\sqrt{4\rho^2 - \nu^2 t}\right)\right\}^2, \tag{48}$$

$$D_\pm(t) = \frac{1}{4} + \frac{2\rho^2}{4\rho^2 - \nu^2}\left\{\sinh\left(\sqrt{4\rho^2 - \nu^2 t}\right)\right\}^2 \mp \tag{49}$$

$$\frac{\rho\sinh\left(\sqrt{4\rho^2-\nu^2}t\right)}{(4\rho^2-\nu^2)}\left\{\sqrt{4\rho^2-\nu^2}\cosh\left(\sqrt{4\rho^2-\nu^2}t\right)\sin 2\nu t+\nu\sinh\left(\sqrt{4\rho^2-\nu^2}t\right)\cos 2\nu t\right\},$$

As it follows from relations (46), (48), at $|\nu| > 2\rho$ the population of the quantized mode oscillates and its amplitude can take arbitrary large value, at $|\nu| < 2\rho$ the population increases exponentially with time. But here we have to take into account the restriction (4). In other words, these results are valid only for the temporal interval, for which the restriction (4) is valid. The behavior of $D_+(t)$ at different values of parameter $\zeta = |\rho/\nu|$ is plotted in figures 1-4. One can see that for some part of time the quantized radiation is squeezed ($D_+(t) < 0.25$) and when $\zeta \nearrow 1/2$ and $\zeta \searrow 1/2$ the maximum degree of squeezing increases right up to 100%. The behavior of the value $D_-(t)$ is similar to the behavior of the value $D_+(t)$.

5.3 Special case

A specific situation arises at $\nu = 2\varepsilon\rho, \varepsilon = \pm 1$. This case can be considered with the help of the limit process $\zeta \to 1/2$. But here we prefer the more direct way. In this case we have, repeating the preceding considerations,

$$\varphi(z,t) = \exp\left[\frac{\varepsilon z^2\exp(2i\nu t)}{2}\right]\int \exp\left[izs\exp(i\nu t)+i\rho ts^2-\frac{i\nu t}{2}\right]\Pi(s)ds,$$

where function $\Pi(s)$ should be determined from the initial conditions. If the initial state is coherent, $\varphi(z,0) = \exp(\alpha z)$,

$$\varphi(z,t) = \frac{1}{\sqrt{1-i\nu t}}\exp\left\{\frac{2\alpha z\exp(i\nu t)-i\nu tz^2\exp(2i\nu t)-i\nu\alpha^2 t}{2(1-i\nu t)}-\frac{i\nu t}{2}\right\}$$

(here we suppose for the distinctness $\varepsilon = 1$). Some algebra leads to the final results at $\alpha = 0$:

$$N(t) = \nu^2 t^2,$$

$$D_\pm(t) = \frac{1}{4}+\frac{\nu^2 t^2}{2}\mp\left[\nu^2 t^2\cos(2\nu t)+\nu t\sin(2\nu t)\right]/2. \quad (50)$$

Note, that the population of the quantized mode increases in this case infinitely. This means that our results are valid for times when our main condition (4) is valid. The behavior of function $D_+(t)$ is plotted on figure 5. One can see, that in this case the degree of squeezing can be up to the 100%. As it follows from the numerical calculations, functions (47) and (49) tend to functions (50) as $\zeta \to 1/2$.

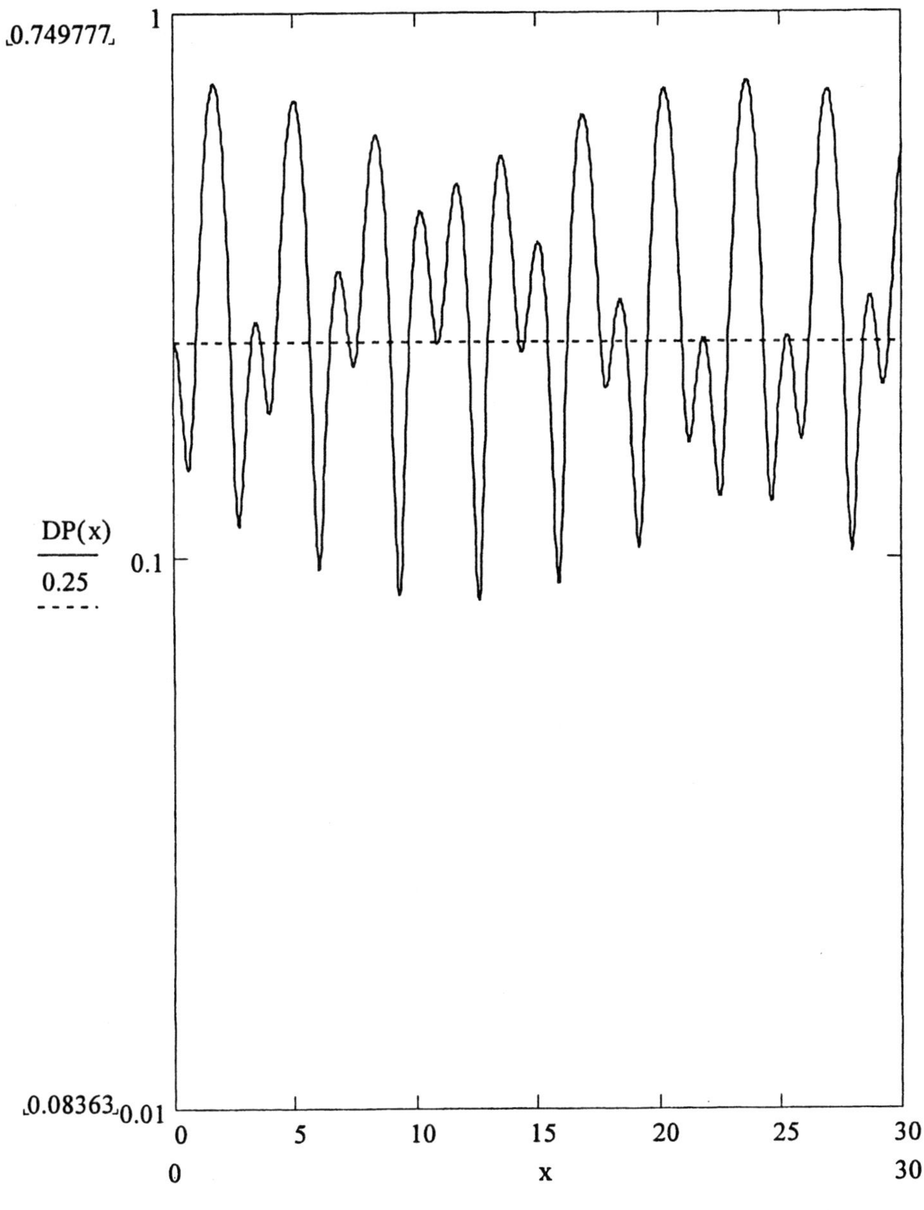

Figure 1: $DP(x) = D_+(x)$ at $\varsigma = 0.25$, $x = \nu t$.

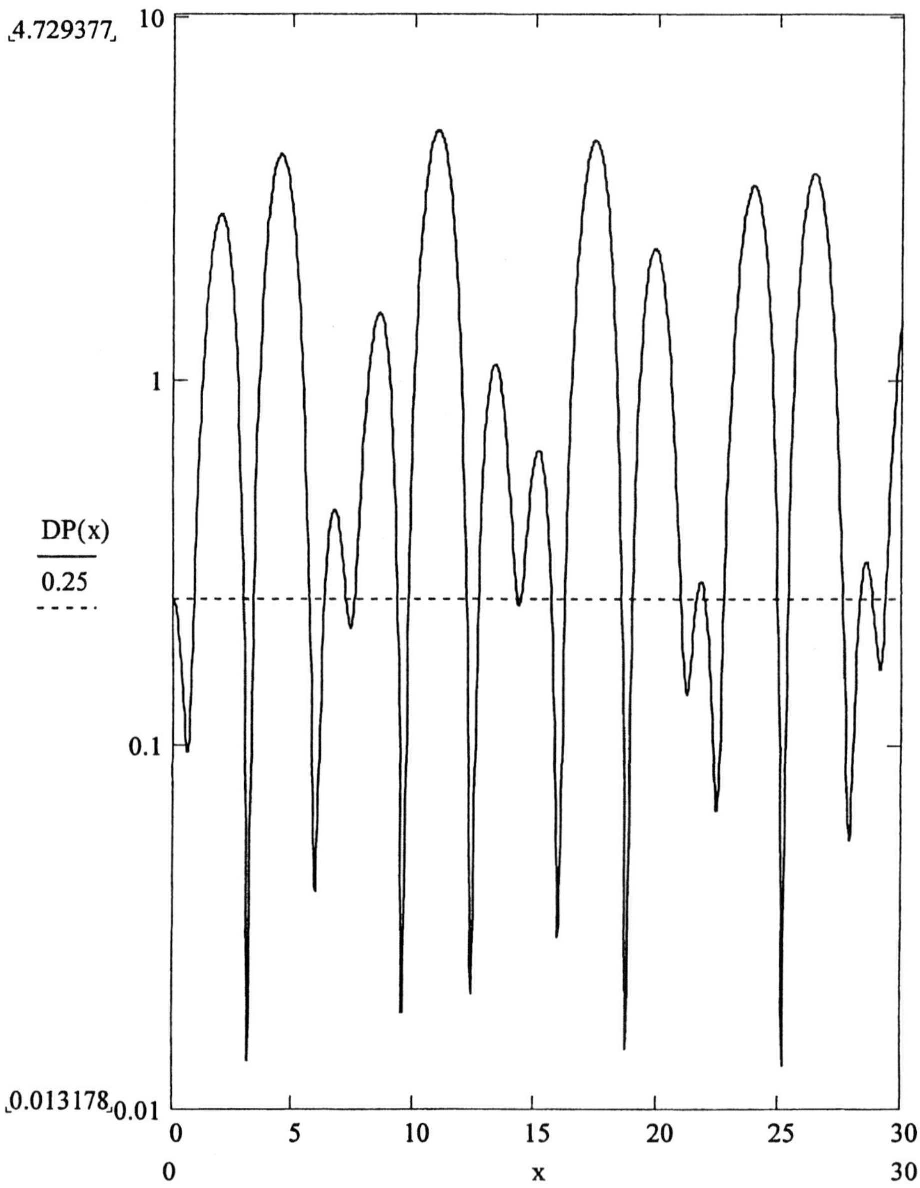

Figure 2: $DP(x) = D_+(x)$ at $\varsigma = 0.45$, $x = \nu t$

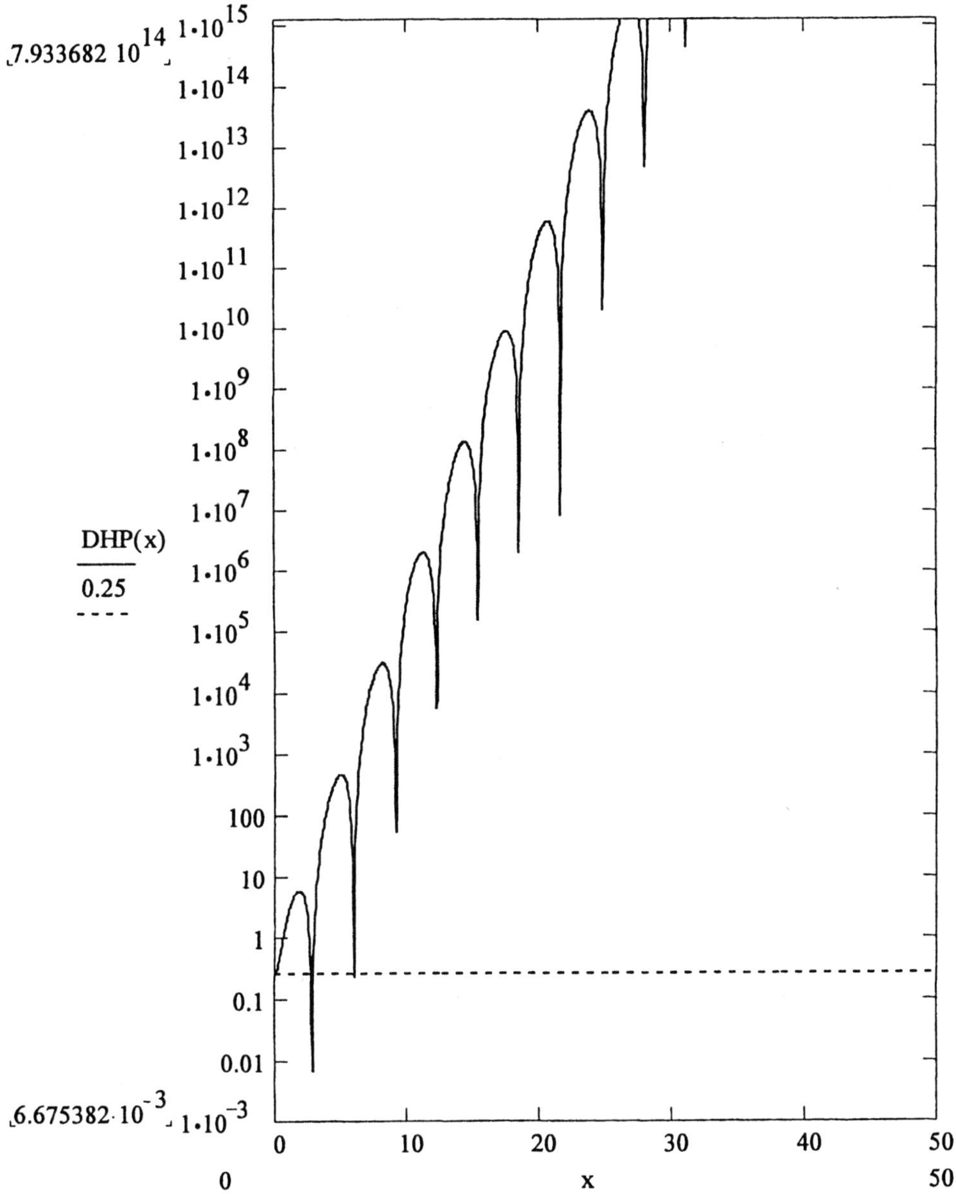

Figure 3: $DHP(x) = D_{+}(x)$ at $\varsigma = 0.6$, $x = \nu t$

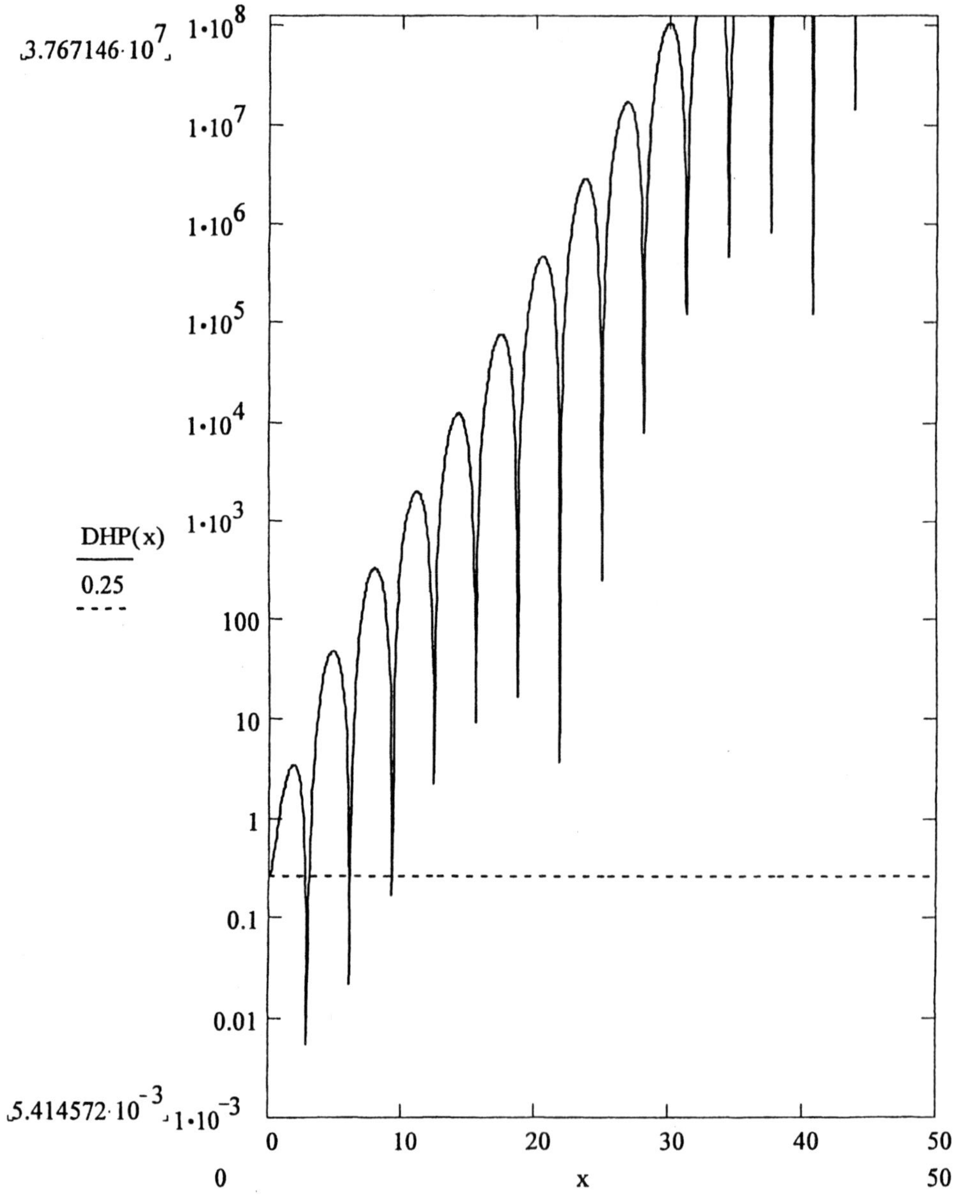

Figure 4: $DHP(x) = D_{+}(x)$ at $\varsigma = 0.52$, $x = \nu t$

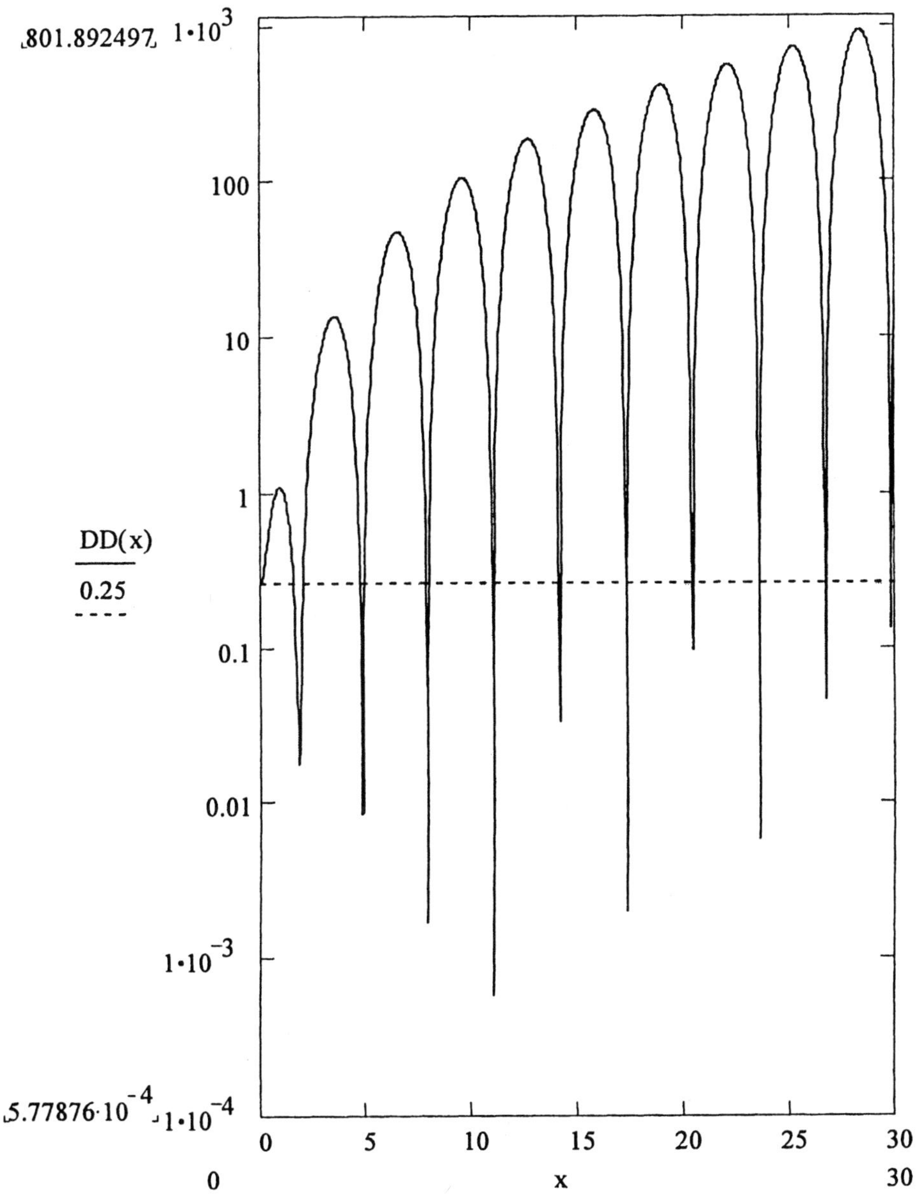

Figure 5: $DD(x) = D_+(x)$ at $\varsigma = 0.5$, $x = \nu t$

6 Summary

We have discussed here the Jaynes-Cummings models interacting with quasiresonant classical fields. We have derived the averaged Hamiltonians for the slow dynamics of our quantum systems. Let us stress once more that our main condition $R >> R_q$ is rather natural, if the coupling constants of the atom's interactions with different modes have equal order of magnitude. We discuss here the two-level atom. Evidently, these results can be modified for the case of the three-level atom too (see, for instance, [25]). From the formal point of view, the averaging procedure discussed above produces the "simple" Hamiltonians from the more complex ones. As it was shown above, in the case $| K\omega - \Omega |= 2 | \nu |<< R$, where ω and Ω are the frequencies of the quantized and the classical radiation respectively, R is the Rabi parameter of the classical field, this procedure leads to the averaged Hamiltonians, which are the products of Fock operators and some matrix operators. In the basis of the eigenvectors of the matrix operator the initial value problem for the averaged Hamiltonians can be solved in explicit form for the one-photon ($K = 1$) and the two-photon ($K = 2$) interactions. This fact gives the possibility to calculate the main quantum-statistical characteristics of the quantized radiation in the cavity. Specifically, our formalism describes the generation of squeezed radiation for the two-photon interaction completely analytically. Moreover, it gives the possibility to examine the collective behavior of the set of N atoms interacting with the quantized and the classical modes simultaneously. The situation of the one-photon interaction was discussed above, and the cooperative effect was discovered. The case of the two-photon interaction we hope to consider elsewhere.

Note that there are no conservation of the excitations in the physical systems considered here - neither in a system governed by Hamiltonian (1), nor in the systems governed by our averaged Hamiltonians. The source of the photons in the quantized mode is the classical fields, and the atom takes part in this process as a transmitter of photons between the classical and the quantized modes.

The author expresses his gratitude to S.V.Siparov for valuable discussions.

References

[1] E.T.Jaynes, F.W.Cummings. Proc. IEEE, 1963, v.51, 89.

[2] P.Alsing, M.S.Zubairy. J.Opt.Soc.Am. B, 1987, v.4, 177.

[3] Walther H. Uspekhi Fiz.Nauk, 1996, v.166, N7, p.777 (in Russian).

[4] S.J.Phoenix, P.L.Knight. JOSA B, 1990, v.7, p.116.

[5] T.Nasreen, K.Zaheer. Phys.Rev.A, 1994, v.49, p.616.

[6] C.K.Law, J.H.Eberly. Phys.Rev.A, 1991,v.43, p.6337.

[7] P.Alsing, D.-S.Guo, H.J.Carmichael. Phys.Rev.A, 1992, v.45, p.5135.

[8] I.V.Jyotsna, G.S.Agarwal. Opt. Commun., 1993, v.99, p. 344.

[9] M.Kozierowski. Phys. Rev. A, 1993, v.47, 723.

[10] Y.-T. Chough, H.J.Carmichael. Phys.Rev.A, 1996, v.54, 1709.

[11] F.-L. Li, S.-Y.Gao. Phys. Rev. A, 2000, v.62, N 043809.

[12] A.Joshi. Phys.Rev.A, 2000, v.62, N 043812.

[13] A.H.Nayfeh. Introduction to Perturbation Techniques. J.Wiley and Sons, N.Y., 1981.

[14] S.M.Chumakov, S.M.Klimov, J.J.Sanchez-Mondragon. Phys. Rev.A, 1994, v.49, p.4972

[15] A.M.Perelomov. Generalized coherent states and theirs applications. Springer Verlag, New York, 1986.

[16] E.I.Aliskenderov, A.S.Shumovsky, Ho Trung Dung. Phys. Part. Nuclei, v.24, 177 (1993).

[17] G.I.Toptygina, E.E.Fradkin. JETP, 1990, v.69, p.428.

[18] A.Ya.Kazakov. JETP, 1993, v.75, p.755.

[19] M.Z. Smirnov. JOSA B,1994, v.11, p.109.

[20] A.Ya.Kazakov. Quant.Semicl.Opt., 1998, 10, 753.

[21] A.Ya.Kazakov. Theor.Math.Phys., 1998, 117, 92.

[22] R.A.Ismailov, A.Ya.Kazakov. JETP, 1999, v.89, 484.

[23] R.A.Ismailov, A.Ya.Kazakov. JETP, 2000, v.91, 691.

[24] A.V.Andreev, V.I.Emel'yanov, Yu.A.Il'inskii. Cooperative effects in optics. IOPP, Bristol, 1993.

[25] A.Ya.Kazakov. Journ.Opt. B, 2001, v.3, N3, 97 (in press).

7 Appendices

7.1 Averaging procedure

We will discuss here the procedure used above for finding the leading asymptotic term of the solution of the initial problem. Let us discuss the initial problem

$$\frac{d\Phi(\tau)}{d\tau} = \varepsilon M(\tau)\Phi(\tau), \tag{51}$$

$$\Phi(0) = Z. \tag{52}$$

Here $\Phi(\tau)$ is a vector function, Z is a vector of the same dimension; we suppose, that $\varepsilon << 1$, matrix - valued function $M(\tau)$ and its derivatives have values of unity. Let us introduce the scaling $\tau_1 = \varepsilon\tau$. We will solve (51), (52) in the form:

$$\Phi(\tau) = \Phi(\tau, \tau_1) = \Phi_0(\tau, \tau_1) + \varepsilon\Phi_1(\tau, \tau_1) + ... \tag{53}$$

Substituting (53) into (51) we get a recurrent system of equations for the functions $\Phi_k(\tau, \tau_1)$. The first equation reads:

$$\frac{\partial\Phi_0(\tau, \tau_1)}{\partial\tau} = 0.$$

Hence,

$$\Phi_0(\tau, \tau_1) = \Phi_0(\tau_1).$$

The second equation reads:

$$\frac{\partial\Phi_0(\tau_1)}{\partial\tau_1} + \frac{\partial\Phi_1(\tau, \tau_1)}{\partial\tau} = M(\tau)\Phi_0(\tau_1). \tag{54}$$

We split this equation into two equations, collecting the fast and slow terms separately. (We call a function fast (slow) if it depends on τ (τ_1)). As a result we have:

$$\frac{\partial\Phi_0(\tau_1)}{\partial\tau_1} = \langle M(\tau)\rangle\, \Phi_0(\tau_1), \tag{55}$$

$$\frac{\partial\Phi_1(\tau, \tau_1)}{\partial\tau} = [M(\tau) - \langle M(\tau)\rangle]\, \Phi_0(\tau_1). \tag{56}$$

Here $\langle ... \rangle$ means that we have omitted the fast terms. Take into account condition (52) and equation (55) we can find $\Phi_0(\tau_1)$. It follows from (56):

$$\Phi_1(\tau, \tau_1) = \Phi_0(\tau_1) \int_0^\tau [M(s) - \langle M(s)\rangle]\, ds + Q(\tau_1),$$

and

$$Q(0) = 0.$$

Continuing this procedure we can construct the full asymptotic expansion of the solution of the problem (51), (52). The analysis of this expansion leads to the conclusion that $\Phi_0(\tau_1)$ is the leading asymptotic term of the expansion up to the time $\tau_1 \sim \varepsilon^{-\theta}$, $0 < \theta < 1$, in other words, $\tau \sim \varepsilon^{-1-\theta}$.

Generally speaking, the use of this technique in problems of quantum optics is not trivial. The reason for this is the fact that operator $M(t)$ can be unbounded and the usual derivation of the asymptotic estimations like [15] is not applicable. But if we can estimate for the initial state the values of the $M(t)$ and its iterations (and this is valid when we discuss the excitation of the quantized mode), it is possible to derive an appropriate version of the perturbation theory. We will discuss this problem elsewhere.

7.2 General solution of the partial differential equation

We consider here the general solution of equation (40). First of all let us substitute

$$\varphi(y,\tau) = \exp\left[\beta y^2/2\right]\Theta(\tau,y), \tag{57}$$

in order to eliminate the term y^2 in equation (40). Then the equation for function $\Theta(\tau, y)$ reads:

$$i\Theta_\tau = \rho\left[\Theta_{yy} + 2\kappa y\Theta_y + (\beta + \beta^2 y^2 + y^2)\Theta\right] + \nu y\left[\beta y\Theta + \Theta_y\right].$$

If we choose

$$\beta = -\frac{\nu + \varepsilon\sqrt{\nu^2 - 4\rho^2}}{2\rho}, \varepsilon = \pm 1, \tag{58}$$

the term proportional to y^2 can be eliminated. The choice of the value ε does not influence the final results and we can suppose $\varepsilon = 1$. Then

$$i\Theta_\tau - \rho\left[\Theta_{yy} + \beta\Theta\right] - (\nu + 2\beta\rho)\, y\Theta_y = 0. \tag{59}$$

A solution of the last equation we seek in the form:

$$\Theta(y,\tau) = \int \exp\left[iys + i\rho\tau s^2\right] q(\tau, s)ds, \tag{60}$$

$$L = \int \exp\left[iys + i\rho\tau s^2\right]\left[iq_\tau - i\left(\nu + 2\beta\rho\right) syq - \beta\rho q\right] ds = 0.$$

Note that $iy\exp\left[iys\right] = d\exp\left[iys\right]/ds$. Integration by parts leads to the relation:

$$L = \int ds \exp\left[iys + i\rho\tau s^2\right] \left\{iq_\tau + (\nu + 2\beta\rho)\, sq_s + 2i\,(\nu + 2\beta\rho)\, \rho\tau s^2 q + (\nu + \beta\rho)\, q\right\} = 0.$$

So, if $q(\tau, s)$ is a solution of the equation

$$iq_\tau + (\nu + 2\beta\rho)\, sq_s + 2i\,(\nu + 2\beta\rho)\, \rho\tau s^2 q + (\nu + \beta\rho)\, q = 0, \tag{61}$$

then function $\Phi(\tau, y)$, defined by (60), is a solution of equation (59) (with some restrictions due to integration by parts). This is an equation in partial derivatives of the first order. We can solve it by separation of variables. After the change of variables

$$s = \exp\left[-i\,(\nu + 2\beta\rho)\,\theta\right] x, \tau = \theta, \tag{62}$$

we obtain:

$$q_\theta + \left[2\rho\,(\nu + 2\beta\rho)\, x^2\theta \exp\left[-2i\,(\nu + 2\beta\rho)\,\theta\right] - i\,(\nu + \beta\rho)\right] q = 0.$$

Then

$$q(\theta, x) = \exp\left[i\,(\nu + \beta\rho)\,\theta - 2\rho\,(\nu + 2\beta\rho)\, x^2 R(\theta)\right] \Pi(x),$$

where $\Pi(x)$ is any function, and

$$R(\theta) = \exp\left[-2i\,(\nu + 2\beta\rho)\,\theta\right] \left[\frac{i\theta}{2\,(\nu + 2\beta\rho)} + \frac{1}{4\,(\nu + 2\beta\rho)^2}\right],$$

where integration is realized along a suitable contour. After substitution into equation (60), we get:

$$\Theta(\tau, y) = \int ds \exp\left[iys + i\,(\nu + \beta\rho)\,\tau - \frac{\rho s^2}{2\,(\nu + 2\beta\rho)}\right] \Pi(s \exp\left[i\,(\nu + 2\beta\rho)\,\tau\right]).$$

Returning to the initial variables, we get relation (41).

Theoretical Physics 2002, Part 1
Edited by Thomas F. George and Henk F. Arnoldus, pp. 57-73
ISBN 1-59033-435-3

Effect of Negative and Classically-Forbidden Momenta in One-Dimensional Quantum Scattering

A. D. Baute[†], I. L. Egusquiza* and J. G. Muga[†]

Abstract

The transmitted wave that results from a collision of a wave packet which is initially to the left of a potential barrier depends in general on the amplitudes of negative momenta of the initial state. The exact form of this dependence is shown and the importance of this classically forbidden effect is illustrated with numerical examples. Special care is taken to account properly for bound states.

I. INTRODUCTION

Suppose that a classical ensemble of independent particles in one dimension is initially confined (at $t = 0$) in the spatial interval $a < x < b \leq 0$, and allowed to move freely after $t = 0$. Only particles with positive momenta may arrive at positive positions for $t > 0$. In contrast, the quantum wave function involves negative-momentum contributions as well,

$$\psi(x,t) = h^{-1/2} \int_{-\infty}^{\infty} dp\, e^{ipx/\hbar} \tilde{\psi}(p) e^{-iE_p t/\hbar}, \tag{1}$$

where $E_p = p^2/2m$, and

$$\tilde{\psi}(p) = h^{-1/2} \int_{-\infty}^{\infty} dx\, e^{-ipx/\hbar} \psi(x,0) \tag{2}$$

---

*Fisika Teorikoaren Saila, Euskal Herriko Unibertsitatea, 644 Posta Kutxa, 48080 Bilbao, Spain

[†]Departamento de Química-Física, Universidad del País Vasco, Apartado 644, 48080 Bilbao, Spain

is the momentum representation of the initial state. The effect of negative momenta for $x, t > 0$ is however a transient one; the total final probability to find the free particle at $x > 0$ is given only by positive momentum components,

$$P_T(\infty) \equiv \lim_{t\to\infty} \int_0^\infty dx\, |\psi(x,t)|^2 = \int_0^\infty dp\, |\tilde{\psi}(p)|^2, \tag{3}$$

see e.g. [1–3]. This negative-momentum effect is also present in collisions, where it is combined with other classically forbidden effects. Consider the family of cut-off potentials of the form

$$V(x) = \begin{cases} 0 & \text{if } x < c, \\ U(x), & \text{if } c \le x \le d, \\ V_0, & \text{if } x > d, \end{cases} \tag{4}$$

where $V_0 \ge 0$ and $U(x)$ are real, as depicted in Fig. 1.

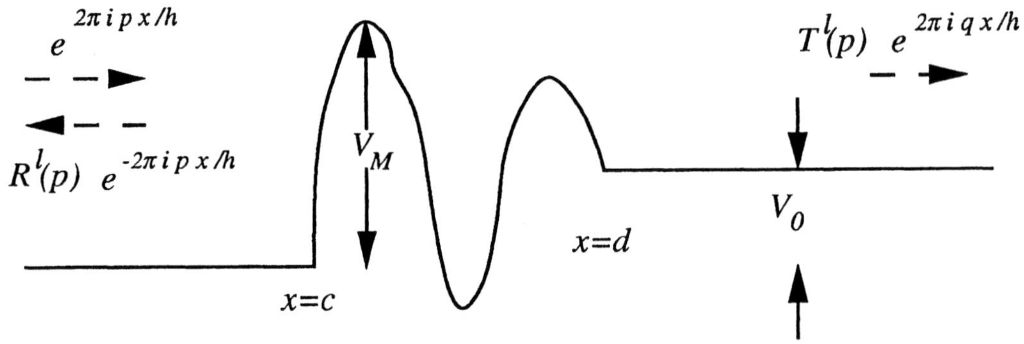

Figure 1. Scattering process in a "cut-off" potential: the support of the potential is compact; an incident wave with momentum p produces a transmitted and a reflected wave.

Let us suppose that the maximum value of the potential is V_M. For a classical ensemble confined between a and b (such that $a < b \le c \le d \le 0$) only particles with *initial* momentum above the "barrier momentum" $p_M \equiv (2mV_M)^{1/2}$ may pass to the right of the potential region. In the quantum case however, there are contributions from all the components of the initial wave packet: (a) $p > p_M \equiv (2mV_M)^{1/2}$ (above-the-barrier transmission); (b) $p_0 < p < p_M$ ("asymptotic" tunneling, these momenta contribute to the transmission probability $P_T(\infty)$), where $p_0 \equiv (2mV_0)^{1/2}$; (c) $0 < p < p_0$ ("transient tunnelling", these momenta do not contribute to $P_T(\infty)$); (d) $-\infty < p < 0$ (transient negative-momentum effect); (e) $p_j = i\gamma_j$, $\gamma_j > 0$ (tunnel effect associated with bound states, which do contribute to $P_T(\infty)$).

While the various tunnel effects have been well discussed in the literature, even though not as clearly distinguished as in the above classification, the negative momentum effect, actually the only one that survives for free motion, has been frequently overlooked, a clear exception being [4,5]. This paper is complementary to the more usual treatments of one dimensional scattering [6–9], and lies in the wake of [4,5], where all these effects were considered in a concise manner, including bound states. Our S-matrix treatment is more explicit than that of [4,5], and shows how the resolution of the identity in scattering eigenstates leads to compact expressions for the transmitted wave packet. We also illustrate the negative momentum effect with some numerical examples and work out in detail both the case when the state is confined initially to the lower level and when the initial confining is to the upper level (for which case the frequently overlooked contributions are those of positive momenta, including evanescent waves). The contribution of bound states is indicated explicitly, both formally and with some numerical examples.

II. SCATTERING EIGENSTATES OF THE HAMILTONIAN

The total Hamiltonian $H = H_0 + V$ may have a discrete set of bound states $\{|E_j\rangle\}$, with energies $E_j < 0$ and (real) wavefunctions $\phi_j(x)$, and a continuum of "stationary scattering eigenstates" with $E_p > 0$. Only the former belong to the Hilbert space of square integrable functions. The latter however form a convenient basis normalized according to Dirac's delta. For energies above V_0, the energy spectrum is doubly degenerate as corresponds physically to incidence from one side or the other. Below V_0 there is only one linearly independent solution. The resolution of the identity may be written in different ways, in particular as

$$1 = \sum_j |E_j\rangle\langle E_j| + \int_{-\infty}^{-p_0} dp\, |p^\pm\rangle\langle p^\pm| + \int_{p_0}^{\infty} dp\, |p^\pm\rangle\langle p^\pm| \pm \int_0^{\pm p_0} dp\, |p^\pm\rangle\langle p^\pm|, \quad (5)$$

where the states $|p^\pm\rangle$ have energy $E_p = p^2/(2m)$. Continuum and bound states are orthogonal. Moreover, $\langle p^\pm|p'^\pm\rangle = \delta(p - p')$, and $\langle E_j|E'_j\rangle = \delta_{jj'}$. The first two integrals in Eq. (5) reflect the double degeneracy signalled above, whereas the last integral corresponds to the non-degenerate part of the continuous spectrum. The reason for the sign choice in front of the last integral is due to the unified notation we shall now introduce for the generalized eigenstates associated with the continuous part of the spectrum.

The states $|p^+\rangle$ with $p > 0$ are characterized by an *incident* plane wave of momentum p from the left. The asymptotic behaviour of their wave-functions $\psi_{p^+}(x)$ is

$$\psi_{p^+}(x) = \frac{1}{h^{1/2}} \times \begin{cases} \exp(ipx/\hbar) + R^l(p)\exp(-ipx/\hbar), & x < c \\ T^l(p)\exp(iqx/\hbar), & x > d, \end{cases} \tag{6}$$

where q is the momentum with respect to the upper (right) level of the potential, $q = (p^2 - 2mV_0)^{1/2}$. For the evanescent regime, i.e. when $p^2 < 2mV_0$, the plane waves on the right become decaying exponentials so the positive imaginary square root is taken. $R^l(p)$ and $T^l(p)$ are reflection and transmission amplitudes for left incidence. They are obtained by solving the stationary Schrödinger equation subject to the boundary conditions specified in (6).

The states $|p^+\rangle$ for $p < -p_0$ are defined by having an *incident* plane wave from the right, oscillating with spatial frequency $|q|/\hbar$, where now the negative square root is taken, $q = -(|p^2 - 2mV_0|)^{1/2}$. The asymptotic behaviour of the corresponding wave-functions is

$$\psi_{p^+}(x) = \frac{1}{h^{1/2}}\left(\frac{p}{q}\right)^{1/2} \times \begin{cases} T^r(-p)\exp(ipx/\hbar), & x < c \\ \exp(iqx/\hbar) + R^r(-p)\exp(-iqx/\hbar), & x > d, \end{cases} \tag{7}$$

The factor $(p/q)^{1/2}$ is necessary for the proper delta normalization. $R^r(-p)$ and $T^r(-p)$ are reflection and transmission amplitudes for right incidence. Note that the arguments of transmission or reflection amplitudes are always positive for states $|p^+\rangle$, independently of the sign of p.

In both cases, (6) and (7), q may be defined as the square root of $p^2 - 2mV_0$ with a branch cut that joins the branch points $p = \pm p_0$ going slightly below $\mathrm{Im}(p) = 0$. In this way the sign of q is the same as the sign of p for $p^2 > p_0^2$.

The states $|p^-\rangle$ are obtained simply by considering $p < 0$ in expression (6) and $p > p_0$ in expression (7). In these states there appears an outgoing plane wave instead of an incident plane wave. Thus, the right hand side of Eq. (6), with $p < 0$, defines the corresponding states $|p^-\rangle$, as those with a left-outgoing plane wave of absolute momentum $|p|$, and can be read as $\psi_{p^-}(x)$, for $p < 0$. For $p < -p_0$, $q < 0$, whereas for $-p_0 < p < 0$, q becomes as before a positive imaginary number. States $|p^-\rangle$ with a right-outgoing plane wave are defined by the right hand side of Eq. (7), with $p > p_0$, and $q > 0$. Note that the arguments of the amplitudes $R^{r,l}(p)$ and $T^{r,l}(p)$ are negative for $|p^-\rangle$ states. Since the formal boundary conditions that define the states $|p^\pm\rangle$ are in fact equal, the negative-argument amplitudes will be given by the same formal expressions valid for their positive-argument counterparts. For the same reason, we shall refer to $R^{r,l}(p)$ and $T^{r,l}(p)$ as "reflection" and "transmission" amplitudes independently of the sign of p, even though, on physical grounds, this terminology and notation would only be appropriate for the $p > 0$ case (only in that case does $T^{r,l}$ multiply a *transmitted* wave and $R^{r,l}$ a *reflected* wave).

Reflection and transmission amplitudes are not independent. The unitarity of the S matrix imposes certain relations among them. The S matrix elements are defined as the coefficients multiplying the outgoing plane waves when the incident plane wave is normalized to unit flux. When the two channels of the one dimensional scattering are open (this happens for $p > p_0$), they are given by

$$\mathsf{S}(p) = \begin{pmatrix} \left(\frac{q}{p}\right)^{1/2} T^l(p) & R^l(p) \\ R^r(p) & \left(\frac{p}{q}\right)^{1/2} T^r(p) \end{pmatrix} \tag{8}$$

The unitarity of the S matrix, $\mathsf{S}\mathsf{S}^\dagger = \mathbf{1}$, implies that

$$\frac{p}{q}|T^r(p)|^2 + |R^r(p)|^2 = 1\,, \tag{9}$$

$$|R^l(p)|^2 + \frac{q}{p}|T^l(p)| = 1\,, \tag{10}$$

$$\frac{p}{q}T^r(p)R^l(p)^* + R^r(p)T^l(p)^* = 0\,. \tag{11}$$

For $0 < p < p_0$ only one channel is open, and the S matrix reduces to a number, $R^l(p)$. Unitarity implies in this case

$$R^l(p)R^l(p)^* = 1, \qquad 0 < p < p_0\,. \tag{12}$$

All these equations, from (9) to (12), are also valid for negative momenta and relate the amplitudes associated with $|p^-\rangle$ states. They can also be derived by comparing various Wronskians of the stationary scattering states at different regions.

For the evanescent case, $0 < p < p_0$, one important relation follows by multiplying $|p^+\rangle$ by $R^l(p)^*$ and using (12). This gives the state $|-p^-\rangle$. Equating the coefficients for $x > d$,

$$T^l(-p) = T^l(p)R^l(p)^*, \qquad 0 < p < p_0. \tag{13}$$

Similarly, by taking the complex conjugate of the boundary conditions (6) and (7), it is found that $\psi_{-p^-}(x) = \psi_{p^+}(x)^*$, and comparing the coefficients that multiply the exponentials,

$$T^{r,l}(-p) = T^{r,l}(p)^*\,, \tag{14}$$

$$R^{r,l}(-p) = R^{r,l}(p)^*\,. \tag{15}$$

Another important relation between $T^l(p)$ and $T^r(p)$ follows by equating the Wronskians of $\psi_{p^\pm}(x)$ and $\psi_{-p^\pm}(x)$ at $x < c$ and $x > d$,

$$T^r(p)p = T^l(p)q. \tag{16}$$

As an illustration of the above, the reflection and transmission amplitudes for the simple step potential ($c = d = 0$, $U(x) = 0$) are given, for all p, by

$$\begin{aligned} T^l(p) &= \frac{2p}{q+p}; \qquad R^l(p) = \frac{p-q}{q+p}; \\ T^r(p) &= \frac{2q}{p+q}; \qquad R^r(p) = \frac{q-p}{p+q}. \end{aligned} \tag{17}$$

The reader may easily check relations (9-16) for amplitudes that correspond to the step potential, Eq. (17), as a test of their validity.

III. A COMPACT EXPRESSION FOR THE TRANSMITTED WAVE FUNCTION

In this section we shall find, using the relations of the previous one, an expression for $\psi(x,t)$, with $t > 0$ and $x > d$, assuming that the initial wave function is restricted to $a < x < b \le c$. First we insert the resolution of the identity in terms of bound states and scattering states $|p^+\rangle$,

$$\begin{aligned} \psi(x,t) &= \sum_j \phi_j(x)\langle E_j|\psi(0)\rangle e^{-iE_jt/\hbar} \\ &+ \int_{-\infty}^{-p_0} dp\, \psi_{p^+}(x)\langle p^+|\psi(0)\rangle e^{-iE_pt/\hbar} \\ &+ \int_0^{\infty} dp\, \psi_{p^+}(x)\langle p^+|\psi(0)\rangle e^{-iE_pt/\hbar}, \qquad \forall x \text{ and } \forall t. \end{aligned} \tag{18}$$

Because of the initial restriction of the wave function, the matrix element $\langle p^+|\psi(0)\rangle$ may be evaluated with the aid of Eqs. (6) and (7),

$$\langle p^+|\psi(t=0)\rangle = \begin{cases} \left(\frac{p}{q}\right)^{1/2} \tilde{\psi}(p) T^r(-p)^*, & p < -p_0, \\ \tilde{\psi}(p) + \tilde{\psi}(-p) R^l(p)^*, & p > 0. \end{cases} \tag{19}$$

Now we use Eqs. (6-7) and (19) in (18), for $x > d$,

$$\begin{aligned} \psi(x,t) &= \sum_j \phi_j(x)\langle E_j|\psi(0)\rangle e^{-iE_jt/\hbar} \\ &+ h^{-1/2} \int_{-\infty}^{-p_0} dp \frac{p}{q} \left[e^{iqx/\hbar} + R^r(-p)e^{-iqx/\hbar}\right] T^r(-p)^* \tilde{\psi}(p) e^{-iE_pt/\hbar} \\ &+ h^{-1/2} \int_0^{\infty} dp\, T^l(p) e^{iqx/\hbar} \left[\tilde{\psi}(p) + R^l(p)^* \tilde{\psi}(-p)\right] e^{-iE_pt/\hbar}, \qquad x > d, \end{aligned} \tag{20}$$

and reorganize the terms as follows,

$$\begin{aligned}\psi(x,t) &= \sum_j \phi_j(x)\langle E_j|\psi(t=0)\rangle e^{-iE_jt/\hbar} \\ &+ h^{-1/2}\int_{-\infty}^{-p_0} dp\,\frac{p}{q}T^r(-p)^*\tilde{\psi}(p)e^{iqx/\hbar}e^{-iE_pt/\hbar} \\ &+ h^{-1/2}\int_{-\infty}^{-p_0} dp\,\left[\frac{p}{q}R^r(-p)T^r(-p)^* + T^l(-p)R^l(-p)^*\right]\tilde{\psi}(p)e^{-iqx/\hbar}e^{-iE_pt/\hbar} \\ &+ h^{-1/2}\int_{-p_0}^{0} dp\,T^l(-p)R^l(-p)^*\tilde{\psi}(p)e^{iqx/\hbar}e^{-iE_pt/\hbar} \\ &+ h^{-1/2}\int_{0}^{\infty} dp\,T^l(p)\tilde{\psi}(p)e^{iqx/\hbar}e^{-iE_pt/\hbar}, \qquad x > d. \end{aligned} \tag{21}$$

Keep in mind that between $-p_0 < p < p_0$, q is a positive imaginary number, $q = i(|p^2 - 2mV_0|)^{1/2}$. The second term in the third line and the fourth line come from a variable change $p \to -p$.

Making use of Eqs. (11-16), there results the simple form

$$\begin{aligned}\psi(x,t) &= \sum_j \phi_j(x)\langle E_j|\psi(t=0)\rangle e^{-iE_jt/\hbar} \\ &+ h^{-1/2}\int_{-\infty}^{\infty} dp\,T^l(p)\tilde{\psi}(p)e^{iqx/\hbar}e^{-iE_pt/\hbar}, \qquad x > d. \end{aligned} \tag{22}$$

The equation can be put in an even more compact form as shown in the next section.

IV. BOUND STATES

None of the particles of the classical ensemble described in the introduction may be trapped by a potential well of the potential $U(x)$. In quantum mechanics though, a bound state wave function extends exponentially beyond the potential limits, and may overlap with the initial state $\psi(0)$, even when this state is localized outside the potential limits. The contribution of these bound states to the wave function at $t > 0$ is orthogonal to the scattering (continuum) part, and will remain spatially linked to the potential region at all times.

A bound state with energy E_j corresponds to a simple pole of $T^l(p)$, or a zero of $1/T^l(p)$ on the positive imaginary axis, at $p_j = i\gamma_j$, $\gamma_j > 0$. In this section we shall see that the bound state terms may be written as a residue

$$\phi_j(x)e^{-iE_jt/\hbar}\langle E_j|\psi(0)\rangle = -2\pi i h^{-1/2}\,\mathrm{Res}\left[T^l(p)\tilde{\psi}(p)e^{iq_jx/\hbar}e^{-iE_pt/\hbar}\right]_{p=p_j}, \qquad x > d, \tag{23}$$

where $q_j = i\sqrt{\gamma_j^2 + p_0^2}$. The function $\tilde{\psi}(p)$ is defined on the complex plane by the integral

$$\tilde{\psi}(p) = \frac{1}{h^{1/2}} \int_a^b dx\, e^{-ipx/\hbar} \psi(x, 0), \tag{24}$$

as has been used all along.

For the formal treatment of bound states it is convenient to introduce Jost solutions $f_1(p, x)$ and $f_2(p, x)$ of the Schrödinger equation. They are defined by the boundary conditions

$$\begin{aligned} f_1(p, x) &\to e^{iqx/\hbar}, & &\text{as } x > d, \quad \text{and} \\ f_2(p, x) &\to e^{-ipx/\hbar}, & &x < c. \end{aligned} \tag{25}$$

We generalize here the treatment of ref. [6] for $V_0 = 0$ to the case $V_0 \geq 0$. It is easy from (6) and (7) to obtain the explicit expressions of f_1 and f_2 at $x < c$ and $x > d$ respectively,

$$f_1(p, x) = \begin{cases} e^{iqx/\hbar}, & x > d, \\ e^{ipx/\hbar} \frac{1}{T^l(p)} + e^{-ipx/\hbar} \frac{R^l(p)}{T^l(p)}; & x < c, \end{cases} \tag{26}$$

$$f_2(p, x) = \begin{cases} e^{iqx/\hbar} \frac{R^r(p)}{T^r(p)} + e^{-iqx/\hbar} \frac{1}{T^r(p)}, & x > d, \\ e^{-ipx/\hbar}. & x < c. \end{cases} \tag{27}$$

In particular, if for some value p_j, $1/T^l(p_j) = 0$, then f_2 and f_1 become real, proportional to each other,

$$f_2 = C f_1, \tag{28}$$

and decay exponentially for $x < c$ and $x > d$. Defining the normalization constant N by

$$N^2 = \int_{-\infty}^{\infty} dx\, f_1^2, \tag{29}$$

we may write the wave-function of the bound state $|E_j\rangle$ as

$$\phi_j(x) = \frac{1}{N} f_1(p_j, x) = \frac{1}{CN} f_2(p_j, x). \tag{30}$$

The overlap between the bound state $|E_j\rangle$ and the initial state, which has support only on the interval $[a, b] \leq c$, is as usual

$$\langle E_j|\psi(0)\rangle = \int_a^b dx\, \phi_j(x)\psi(x,0)\,. \tag{31}$$

On substituting the second equality of Eq. (30) in Eq. (31), and taking into account definition (24) and Eq. (27), we obtain

$$\langle E_j|\psi(0)\rangle = \frac{1}{CN}\int_a^b dx\, f_2(p_j,x)\psi(x,0) = \frac{1}{CN}\int_a^b dx\, e^{-ip_jx/\hbar}\psi(x,0) = \frac{h^{1/2}}{CN}\tilde{\psi}(p_j)\,. \tag{32}$$

Therefore, for $x > d$, we have

$$\begin{aligned}\phi_j(x)\langle E_j|\psi(0)\rangle &= \left(\frac{1}{N}f_1(p_j,x)\right)\langle E_j|\psi(0)\rangle = \left(\frac{1}{N}e^{iq_jx/\hbar}\right)\langle E_j|\psi(0)\rangle \\ &= \frac{h^{1/2}}{CN^2}e^{iq_jx/\hbar}\tilde{\psi}(p_j) = -2\pi i h^{-1/2}e^{iq_jx/\hbar}\tilde{\psi}(p_j)\operatorname{Res}T^l(p)_{p=p_j}\,,\end{aligned} \tag{33}$$

so that (23) is obtained. In the last line of (33) we have used [6]

$$\operatorname{Res}T^l(p)_{p=p_j} = \left[\frac{\partial}{\partial p}(1/T^l(p))\right]^{-1}_{p=p_j} = \frac{i\hbar}{CN^2}, \tag{34}$$

a relation that may be obtained by differentiating the Schrödinger equation for f_1 with respect to p, integrating f_1f_2 between $-R$ and R as $R\to\infty$, and comparing with the derivative of the Wronskian of f_1 and f_2 with respect to p for $p = p_j$, see also [10]. Finally, combining (22) and (23) we can write the transmitted wave packet very compactly as

$$\psi(x,t) = h^{-1/2}\int_\Omega dp\, T^l(p)\tilde{\psi}(p)e^{iqx/\hbar}e^{-iE_pt/\hbar}, \qquad x > d, \tag{35}$$

where the contour Ω goes from $-\infty$ to ∞ passing above the bound state poles, as shown in Fig. 2.

V. INITIAL STATE IN THE UPPER LEVEL

Imagine now that the initial state has support only to the right, where the potential is V_0; that is, assume that the initial support of the state is the interval $[a,b]$, such that $d \le a < b$. In this case, instead of Eq. (19), we have

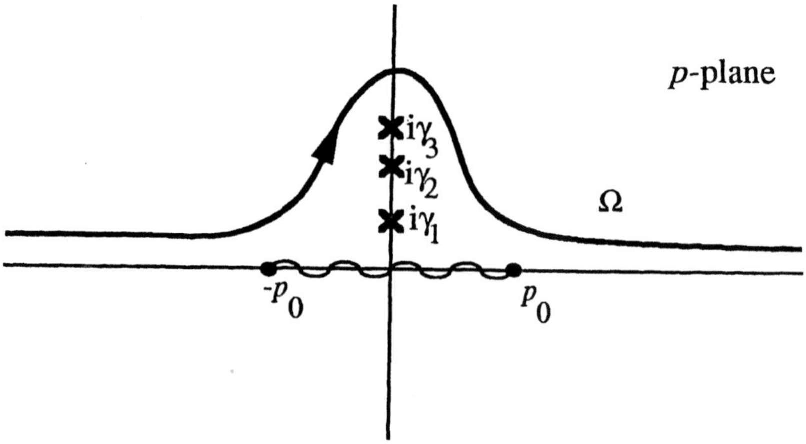

Figure 2. Path of integration for Eq. (35)

$$\langle p^+|\psi(t=0)\rangle = \begin{cases} T^l(p)^*\tilde{\psi}(q)\,, & p > p_0\,, \\ T^l(p)^*\tilde{\psi}(-q)\,, & 0 < p < p_0\,, \\ \left(\frac{p}{q}\right)^{1/2}\left(\tilde{\psi}(q) + R^r(-p)^*\tilde{\psi}(-q)\right)\,, & p < -p_0\,, \end{cases} \tag{36}$$

where, as before, $q = (p^2 - p_0^2)^{1/2}$, with the branch cut going slightly below $\mathrm{Im}(p) = 0$, and $\tilde{\psi}(q)$ is defined by Eq. (24).

Expression (18), as such, is also valid for $x < c$. Let us then substitute Eq. (36) in Eq. (18), and use Eqs. (6) and (7). The result analogous to Eq. (21) reads then, for $x < c$ when the initial state is restricted at $x > d$,

$$\begin{aligned} \psi(x,t) &= \sum_j \phi_j(x)\langle E_j|\psi(t=0)\rangle e^{-iE_jt/\hbar} \\ &+ h^{-1/2}\int_{-\infty}^{-p_0} dp\, T^l(p)^*\tilde{\psi}(q)e^{ipx/\hbar}e^{-iE_pt/\hbar} \\ &+ h^{-1/2}\int_{p_0}^{\infty} dp\, T^l(p)^*\tilde{\psi}(q)e^{ipx/\hbar}e^{-iE_pt/\hbar} \\ &+ h^{-1/2}\int_{0}^{p_0} dp\, T^l(p)^*\tilde{\psi}(-q)e^{ipx/\hbar}e^{-iE_pt/\hbar} \\ &+ h^{-1/2}\int_{p_0}^{\infty} dp\left[\frac{p}{q}R^r(p)^*T^r(p) + T^l(p)^*R^l(p)\right]\tilde{\psi}(q)e^{-ipx/\hbar}e^{-iE_pt/\hbar} \\ &+ h^{-1/2}\int_{0}^{p_0} dp\, T^l(p)^*R^l(p)\tilde{\psi}(-q)e^{-ipx/\hbar}e^{-iE_pt/\hbar}, \qquad x < c. \end{aligned} \tag{37}$$

Making use of the unitarity relations, this expression can be written as

$$\psi(x,t) = \sum_j \phi_j(x)\langle E_j|\psi(t=0)\rangle e^{-iE_jt/\hbar} + h^{-1/2}\int_\Gamma dp\, T^l(-p)\tilde{\psi}(q)e^{ipx/\hbar}e^{-iE_pt/\hbar}\,, \tag{38}$$

where the path of integration Γ goes from $-\infty$ to $+\infty$ immediately *below* the branch cut. Notice that for $p = -i\gamma$, with real positive γ, q becomes $-i\sqrt{\gamma^2+p_0^2}$, which accounts for the minus sign in front of q in the third and fifth integrals of Eq. (37).

In order to include the bound states in a more compact expression, let us substitute the first equality of Eq. (30) in Eq. (31). Thus, using definition (24) and Eq. (26), we obtain

$$\langle E_j|\psi(0)\rangle = \frac{1}{N}\int_a^b dx\, f_1(p_j,x)\psi(x,0) = \frac{1}{N}\int_a^b dx\, e^{iq_jx/\hbar}\psi(x,0) = \frac{h^{1/2}}{N}\tilde{\psi}(-q_j)\,, \tag{39}$$

where, as before, $q_j = i\sqrt{\gamma_j^2+p_0^2}$. Therefore, for $x < c$, we have in this situation, using Eqs. (27) and (30),

$$\begin{aligned}\phi_j(x)\langle E_j|\psi(0)\rangle &= \left(\frac{1}{CN}f_2(p_j,x)\right)\langle E_j|\psi(0)\rangle = \left(\frac{1}{CN}e^{-ip_jx/\hbar}\right)\langle E_j|\psi(0)\rangle \\ &= \frac{h^{1/2}}{CN^2}e^{-ip_jx/\hbar}\tilde{\psi}(-q_j)\,.\end{aligned} \tag{40}$$

The pole of $T^l(p)$ at $p_j = i\gamma_j$ becomes a pole of $T^l(-p)$ at $-p_j$. Thus,

$$T^l(-p) = \frac{-i\hbar}{CN^2}\frac{1}{p+p_j} + \cdots\,, \tag{41}$$

and it follows that

$$\phi_j(x)\langle E_j|\psi(t=0)\rangle e^{-iE_jt/\hbar} = 2\pi i h^{-1/2}\,\mathrm{Res}\left[T^l(-p)e^{ipx/\hbar}\tilde{\psi}(q)e^{-iE_pt/\hbar}\right]_{p=-p_j}\,, \tag{42}$$

from which we can conclude that when the initial state is restricted to the interval $[a,b]$, with $d < a < b$,

$$\psi(x,t) = h^{-1/2}\int_{\Omega'} dp\, T^l(-p)\tilde{\psi}(q)e^{ipx/\hbar}e^{-iE_pt/\hbar}\,, \tag{43}$$

for $x < c$, where Ω' is a path of integration that goes from $-\infty$ to $+\infty$ *below* the branch cut and the poles $-p_j$.

VI. DISCUSSION

The main results of this paper are Eqns. (35) and (43), that provide simple, compact, and exact expressions for the transmitted wave packet in terms of the initial momentum distribution and bound state contributions. Whereas these expressions have been obtained for cut-off potentials, they may be easily generalized for potentials that decay fast enough, so that the initial wave packet does not overlap significantly with the potential region, and for x values where the asymptotic expressions for the scattering states are accurate.

These two equations make clear the need to include different contributions for the transmitted wave-packet. Even though tunnelling terms are of course included in theoretical analysis of wave packet collisions, the contribution of negative momenta is frequently overlooked in integral expressions of the wave function. Similarly, the contribution of the evanescent momentum region when the wave packet is initially on the potential upper level has been also disregarded. We have applied these equations in [11] to establish the relation between source boundary conditions, in which the wave function is specified at a point and for all times, with standard initial-value-problem boundary conditions where the wave function is specified in all space at $t = 0$. One further application is the study of arrival-time measurement models where a clock dial is coupled to the particle's motion in such a way that the particle's crossing stops the dial's motion [12–14]. These models are described by step potentials of the form considered in section V.

Let us now illustrate the formal results obtained above with some simple explicit computations. In all of them, the initial wave packet is taken as the ground state of an infinite well; it is located between a and b,

$$\psi(x,0) = \left(\frac{2}{b-a}\right)^{1/2} \sin[(x-a)k_w]\mathcal{H}(a,b), \tag{44}$$

where $k_w = \pi/(b-a)$, and

$$\mathcal{H}(a,b) = \begin{cases} 1 & a < x < b, \\ 0 & \text{otherwise.} \end{cases} \tag{45}$$

See [11] for details of the analytical time dependence of $\psi(x,t)$. Here we shall evaluate the norm for $x > 0$,

$$P_T(t) = \int_0^\infty dx\, |\psi(x,t)|^2 \,, \tag{46}$$

as well as the contributions from positive and negative momenta, the interference terms, and, whenever required, the contribution of bound states and evanescent waves.

For instance, in the case of free motion, by defining the projectors

$$P = \int_0^\infty dp\,|p\rangle\langle p| \quad \text{and} \tag{47}$$

$$Q = 1 - P\,, \tag{48}$$

P_T is decomposed into three terms, $P_T = P_{T,+} + P_{T,-} + P_{T,int}$:

$$P_{T,+}(t) = \int_0^\infty dx\,|\langle x|P\psi(t)\rangle|^2\,, \tag{49}$$

$$P_{T,-}(t) = \int_0^\infty dx\,|\langle x|Q\psi(t)\rangle|^2\,, \tag{50}$$

$$P_{T,int}(t) = \int_0^\infty dx\,2\mathrm{Re}[\langle x|P\psi(t)\rangle\langle Q\psi(t)|x\rangle]\,. \tag{51}$$

These quantities are shown as functions of t for free motion in Fig. 3. The interference term is important only for early times, and the limit $P_T(\infty) = P_{T,+}(\infty)$ is clearly checked visually, although the decay of $P_{T,-}$ is rather slow.

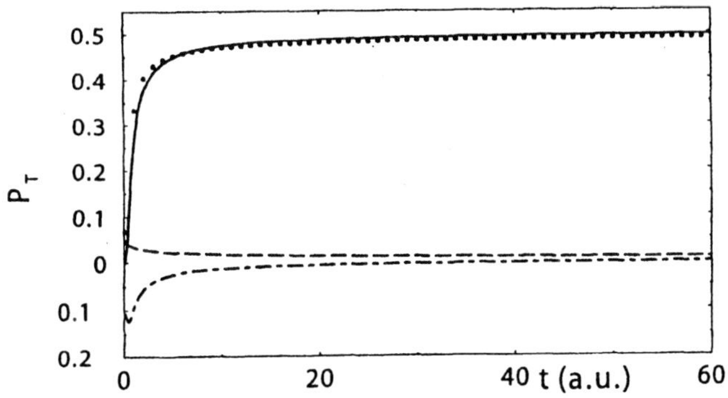

Figure 3. The contributions of positive and negative momentum to the probability of finding the particle in the half line $x > 0$ for time t in the free case are shown with a dotted and a dashed line, respectively. The interference term is depicted with a dash-dotted line, and the total probability of finding the particle in the half line $x > 0$ for time t is portrayed as a thick continuous line. The initial state is given by expression (44), with $a = -2.01$, $b = -0.01$. The mass m equals 1. Atomic units (a.u.) are used throughout.

In Fig. 4 we show how the addition of an attractive delta potential, that creates a bound state $|E_1\rangle$, modifies the analogous contributions, since the bound state component has to be detracted from them. $P_{T,bound}$, given by

$$P_{T,bound}(t) = \int_0^\infty dx\,|\langle x|E_1\rangle\langle E_1|\psi(t)\rangle|^2\,,$$

is in fact constant in time, since there is only one bound state, and it is no longer true that the total final probability $P_T(\infty)$ equals $P_{T,+}(\infty)$; actually, we have $P_T(\infty) = P_{T,bound} + P_{T,+}(\infty)$. In the picture we have only shown the interference term between positive and negative momenta, which is the most relevant one in this case. There are of course two other interference (non-diagonal) terms:

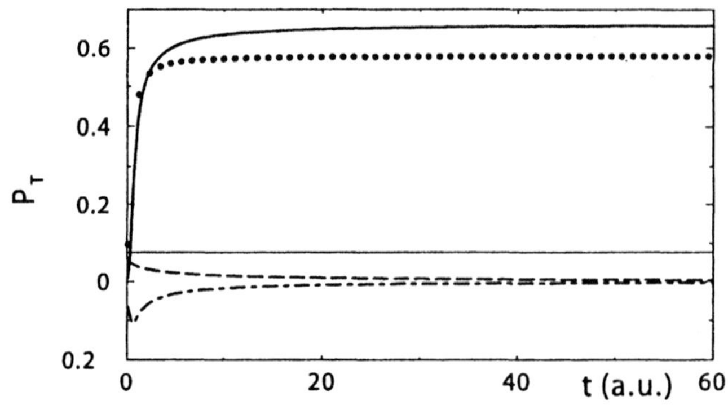

Figure 4. The initial state is as in Eq. (44), but multiplied by $\exp(ipx)$, where $p = 0.5$ (all quantities in atomic units, a.u.). Again we have $a = -2.01$ and $b = -0.01$. The motion of the particle is free everywhere, except for an attractive delta potential, $U(x) = -U_0\delta(x)$, with $U_0 = 1/8$. The contribution of the bound state to the total probability of finding the particle in the half line $x > 0$ is shown with a thin continuous line; the contributions of positive and negative momenta with a dotted and a dashed line, respectively. The interference term between positive and negative momenta is depicted with a dash-dotted line, and the total probability as a thick continuous line.

$$P_{T,b+}(t) = \int_0^\infty dx\, 2\mathrm{Re}[\langle x|E_1\rangle\langle E_1|\psi(t)\rangle\langle\psi(t)P|x\rangle]\,,$$
$$P_{T,b-}(t) = \int_0^\infty dx\, 2\mathrm{Re}[\langle x|E_1\rangle\langle E_1|\psi(t)\rangle\langle\psi(t)Q|x\rangle]\,.$$

In order to show the applicability of expression (35), we portray in Fig. 5 and Fig. 6 the squared moduli of the total amplitude and of different components of the amplitude for a step potential with an attractive delta potential; that is, the potential is $V(x) = -U_0\delta(x) + V_0\theta(x)$. In this situation, (35) may be decomposed into four contributions by drawing the contour Ω very close to the real axis and encircling the bound state: additional to the "positive" ($p > p_0$) and "negative" ($p < -p_0$) momenta components, we have evanescent waves ($|p| < p_0$) and the contribution of the bound state. This leads to ten different terms for the density $|\langle x|\psi(t)\rangle|^2$: four partial densities associated with the four components, and six interference terms, but only the four partial densities are represented in Figs. 5 and 6. At the time of both pictures, ($t = 5$ in atomic units), the evanescent waves are the major contributor to the wavefunction in the vicinity of the origin. It should be noticed though that interference terms are relevant in that region and elsewhere. The magnitude of the component of negative momenta ($p < p_0$), as well as that corresponding to the bound state, is rather small relative to the positive momenta component ($p > p_0$) of the amplitude modulus squared, but it is nonetheless very important even for large distances because of the interference between components, as is clearly reflected in Fig. 5; compare in particular the oscillations of the dotted line ($p > p_0$ contribution) with the smooth solid line of the total density. Fig. 6 is added to show more clearly the relative importance of the negative momenta and bound state components.

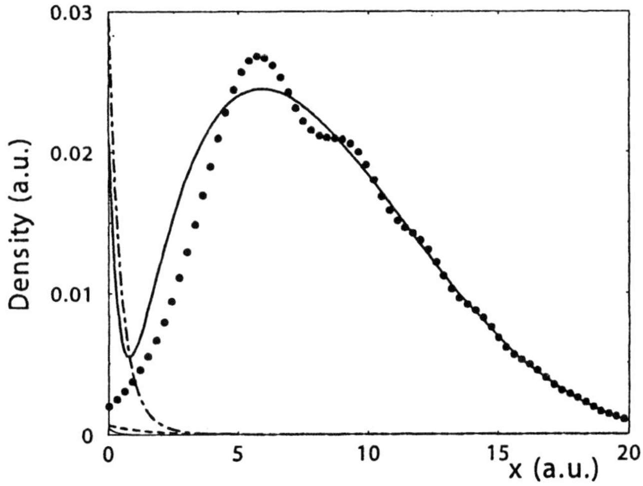

Figure 5. The particle of mass $m = 1$ moves in the potential $V(x) = -U_0\delta(x) + V_0\theta(x)$, where $U_0 = 1/8$ and $V_0 = 1$ (in atomic units). The initial state is as in Eq. (44), with $a = -2.01$ and $b = -0.01$, but multiplied by $\exp(ipx)$, where $p = 0.5$. Time $t = 5$ has elapsed since the initial state was released. The moduli squares of different contributions to the amplitude are depicted: continuous thick line, total amplitude modulus squared; dotted and dashed lines, square modulus of the contribution to the amplitude of "positive" ($p > p_0$) and "negative" ($p < p_0$) momenta; dot-dashed line, evanescent waves ($|p| < p_0$); continuous thin line, bound state contribution.

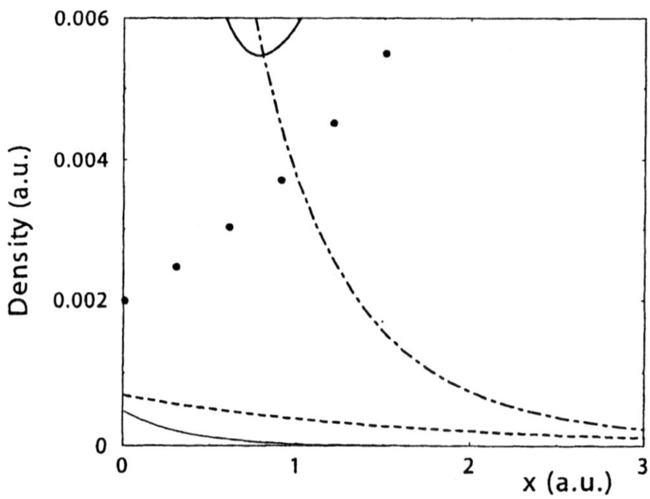

Figure 6. Close-up of Fig. 5 in the lower left corner.

ACKNOWLEDGEMENTS

This paper is dedicated to R. F. Snider on the occasion of his 70th birthday. We acknowledge support by Ministerio de Educación y Cultura (Grants BFM2000-0816-C03-03 and AEN99-0315), by The University of the Basque Country (grant UPV 063.310-EB187/98), and the Basque Government (grant PI-1999-28). A. D. Baute acknowledges an FPI fellowship by Ministerio de Educación y Cultura.

References

[1] W. Jaworski and D. M. Wardlaw, Phys. Rev. A **37**, 2843 (1988).
[2] G. R. Allcock, Ann. Phys. (N.Y.) **53**, 253 (1969).
[3] J. G. Muga, S. Brouard, and R. F. Snider, Phys. Rev. A **46**, 6075 (1992).
[4] C. L. Hammer, T. A. Weber, and V. S. Zidell, Am. J. Phys. **45**, 933 (1977).
[5] T. A. Weber and C. L. Hammer, J. Math. Phys. **18**, 1562 (1977).
[6] R. G. Newton, J. Math. Phys. **21**, 493 (1980).
[7] T. Aktosun, M. Klaus, and C. van der Mee, J. Math. Phys. **37**, 5897 (1996).
[8] M. G. Rozman, P. Reinecker, and R. Tehver, Phys. Rev. **A49**, 3310 (1994).

[9] M. S. Marinov and B. Segev, J. Phys. **A29**, 2839 (1996), quant-ph/9602015.
[10] M. L. Goldberger and K. M. Watson, *Collision Theory* (John Wiley and Sons, 1964).
[11] A. D. Baute, I. L. Egusquiza, and J. G. Muga, Sources of quantum waves, J. Phys. A34, 4289 (2001); quant-ph/0007066.
[12] Y. Aharonov, J. Oppenheim, S. Popescu, B. Reznik, and W. G. Unruh, Phys. Rev. A **57**, 4130 (1998), quant-ph/9709031.
[13] J. G. Muga and C. R. Leavens, Arrival time in quantum mechanics, Phys. Rep., 338, 353 (2000)
[14] A. D. Baute, I. L. Egusquiza, and J. G. Muga, Phys. Rev. A 64, 014101 (2001).

Theoretical Physics 2002, Part 1
Edited by Thomas F. George and Henk F. Arnoldus, pp. 75-81
ISBN 1-59033-435-3

Mapping on a Time-Independent Schrödinger Equation: A Method to Find Analytical Solutions to Generalized Reaction-Diffusion Equations

Mauricio Bellini*†

A recently introduced semi-inverse method is discussed, which allows to find *exact* steady-state solutions for a class of nonlinear equations of which the reaction-diffusion ones are a particular case. The method relies on the fact that—under certain conditions—equations in this class can be mapped on a time-independent Schrödinger one, so exploiting the large collection of exact solutions to the latter. An exact and highly nontrivial outgoing annular wave (or "target") pattern is analytically found as an example, being its features discussed and compared with those of the corresponding reaction-diffusion pattern.

Out of the universe of dissipative nonlinear distributed systems (or "media"), reaction-diffusion (RD) ones have occupied a privileged position during the last decades because of their ability to model many interesting phenomena involving "chemical reactions" in a wide sense (so they apply quite naturally in fields like biology or economics). [3–5] They stand in fact as a paradigm for the study of pattern formation and synchronization phenomena in a transdisciplinary (or "synergetic") approach. [1,2] RD systems are in some sense "ideal", since their interactions with the environment are accounted for by a simple diffusion term. The nonlinearities are *local* and reside in the *direct self-interaction* or "reaction" term, as well as in the possibility that the "diffusion coefficient" depend on the field itself (environment-mediated or *indirect self-interaction*).

Recently, we have introduced a semi-inverse method which allows to obtain analytical solutions to (one-component) systems of this kind by mapping them on a Schrödinger-like equation. We have also worked out several examples ranging from static one-dimensional solutions to travelling annular waves, [6,7] and very recently we have even succeeded in finding spiral solutions. [8] Although the method might not be useful to find a solution to a *specific* RD equation, it does provide analytical forms for the *spatial dependence* of the reaction term F and the diffusion coefficient D, such that the RD pattern has the sought *qualitative* features under a variety of conditions.

Yet more recently, we have extended our proposal to a larger class of equations whose nonlocal term is the gradient of the product of a field-dependent diffusion coefficient times a positive integer power n of the gradient of the field: [9]

$$\frac{\partial}{\partial t}\Phi(\vec{x},t) = \nabla\left\{D[\Phi(\vec{x},t)]\;[\nabla\Phi(\vec{x},t)]^n\right\} + F[\Phi(\vec{x},t)]. \tag{1}$$

Clearly, for $n=1$ we retrieve the ordinary RD equation. As discussed thoroughly in Refs. [6–9], the strategy to find a solution for equations of this kind relies on the observation that a *specific* spatio-temporal dependence $\Phi(\vec{x},t)$ of the field generates *explicit* dependences of both $F(\vec{x},t)$ and $D(\vec{x},t)$ on space and time. Hence, eq.(1) can be rewritten as:

$$\frac{\partial}{\partial t}\Phi(\vec{x},t) = \nabla\left\{D(\vec{x},t)\;[\nabla\Phi(\vec{x},t)]^n\right\} + F(\vec{x},t). \tag{2}$$

However, the latter is not the only motivation to consider nonlinear differential equations like eq. (2): the functions $D(\vec{x},t)$ and $F(\vec{x},t)$ may equally well depend on space and time from

*E-mail: mbellini@mdp.edu.ar
†Instituto de Física y Matemáticas, Universidad Michoacana de San Nicolás de Hidalgo, AP:2-82, (58041) Morelia, Michoacán, México

the outset and not as a consequence of a specific spatio-temporal dependence of the field. In other words the system needs not be autonomous, and the specific forms of $D(\vec{x},t)$ and $F(\vec{x},t)$ may result from spatio-temporal inhomogeneities of the environment. As it is known, nonequilibrium structures can be mantained under these conditions by entropy production. Although we are as yet not aware of *immediate* physical applications of the *specific class* of equations to which we are offering a method of solution, we may regard this restricted class as a member of a richer collection of models *whose evolution equations include powers of the derivatives of the field* (thus accounting for a higher degree of non-locality and nonlinearity in the interactions *between the field and its environment*). Phenomenological models like the "porous media equation", [10] Kuramoto's near-threshold phase equation for type III_o instabilities, [1] or the (deterministic version of the) Kardar-Parisi-Zhang (KPS) equation for surface growth [11] are just known—and simple—examples belonging to this category. Yet several nonlinear maps are known that make important (from the physical viewpoint) equations appear quite differently. The interest of our finding resides in that the possibility that one of those phenomenologically relevant equations be shown to belong to our restricted class cannot be ruled out.

For $n=1$ eq. (2) is a *linear* equation, which we solve by a semi-inverse method:

1. we restrict ourselves to stationary ($\frac{\partial}{\partial t}\Phi^{st}(\vec{x},t)=0$) cases and impose symmetries (which need not be simple, see Ref. [8]), so to deal with ordinary differential equations;
2. for those cases, we map eq.(2) onto a stationary one-dimensional Schrödinger one;
3. finally, we *propose* some functions that determine essentially $\Phi(\vec{x},t)$ and $D(\vec{x},t)$ (so endowing the first with the qualitative features we want the solution to have), and retrieve the functional form of $F(\vec{x},t)$.

In Ref. [9] we have extended this proposal to the case $n>1$, where eq.(2) is not linear anymore. In this work we address the yet less trivial case of outgoing annular waves in two dimensions: $\Phi^{tw}(r,\theta,t)\equiv\phi(r-vt)$, with $v=$ const, or (extrinsic) "target patterns". The term "extrinsic" alludes to the fact that in this case, the *effective* reaction term does not depend *solely* on $\phi(r-vt)$ (see eq.(4) below). In terms of the "comoving " variable $r'\equiv r-vt$, eq.(2) reads:

$$\frac{1}{r'}\frac{d}{dr'}\left[r'\,\tilde{D}(r')\left(\frac{d\phi}{dr'}\right)^n\right]+\tilde{F}(r')+v\,\frac{d\phi}{dr'}=0. \tag{3}$$

As already staded, $\tilde{D}(r')$ and $\tilde{F}(r')$ are shorthands for $D[\phi(r')]$ and $F[\phi(r')]$, respectively. For convenience we shall focus—rather than on $\tilde{D}(r')$ and $\tilde{F}(r')$ themselves—on

$$D(r')\equiv r'\tilde{D}(r') \qquad \text{and}$$
$$F(r')\equiv r'[\tilde{F}(r')+v\,\frac{d\phi}{dr'}]. \tag{4}$$

The *effective* reaction term $F(r')$ acts as an external energy source for the system (which is larger the larger is the propagation speed v).Expanding in eq.(3) the derivative of the expression between square brackets, and dividing by

$$f(r')\equiv n\,D(r')\left(\frac{d\phi}{dr'}\right)^{n-1}$$

(we assume the positiveness of $\phi(r')$ and $D(r')$, and thewell-behavedness of $\frac{d\phi}{dr'}$), equation (3) can be cast in the form:

$$\frac{d^2\phi}{dr'^2} + 2g(r')\left(\frac{d\phi}{dr'}\right) - \alpha(r')\phi(r') = 0,$$

with

$$g(r') \equiv \frac{1}{2n}\frac{d}{dr'}(\ln D)$$

and

$$\alpha(r') \equiv -\frac{F(r')}{f(r')\,\phi(r')}.$$

Then—as in Refs. [6–9]—one sees immediately that the function

$$\psi(r') \equiv [D(r')]^{\frac{1}{2n}}\,\phi(r'),$$

obeys:

$$\frac{d^2\psi(r')}{dr'^2} - G(r')\psi(r') = 0, \tag{5}$$

with $G(r')$ given by

$$G(r') = \alpha(r') + g^2(r') + \frac{dg}{dr'}.$$

Eq.(5) can be regarded (in the appropriate units) as the (steady state) one-dimensional Schrödinger equation for a particle in a potential $U_{Sch}(r')$ with "energy" eigenvalue E:

$$G(r') = U_{Sch}(r') - E.$$

Obviously, we are looking for real and normalizable solutions of eq.(5), i.e. for bound states of $U_{Sch}(r')$. Moreover, if the field $\phi(r')$ represents the concentration of some chemical species or something else akin, it would rather be positive in the whole physical domain $0 < r' < \infty$. For that reason, we shall only concentrate on the ground state $\psi_o(r')$.

As an example we propose

$$G_o(r') = A^2 e^{-2r'} + 2ABe^{-r'} + \left(B - \frac{1}{2}\right)^2,$$

whose corresponding $U_{Sch}(r')$ is a Morse potential (as is better exposed by taking the ground state "energy" eigenvalue to be $E_o = -\left(B - \frac{1}{2}\right)^2$)

$$U_{Sch}(r') = A^2\,e^{-2r'} + 2AB\,e^{-r'},$$

which has its minimum at $r'_o = \ln\left(-\frac{A}{B}\right)$. Since:

$$\psi_o(r') \propto e^{-(B-1/2)r'}\,\exp(Ae^{-r'}),$$

the condition for this potential to have at least one bound state results to be:

$$A < 0; \qquad B > \frac{1}{2}.$$

Let us propose also

$$g(r') = ae^{-r'} + be^{r'},$$

so that the diffusion coefficient is

$$D(r') \propto \exp[2n(-ae^{-r'} + be^{r'})], \tag{6}$$

which is plotted in Fig.1 as a function of $x' = r'\cos\theta$ and $y' = r'\sin\theta$, and is seen to increase without bound as r' does. Having $\psi_o(r')$ and $D(r')$ we obtain the functional form of the concentration pattern:

$$\phi_o(r') \propto e^{(ae^{-r'} - be^{r'})}\, e^{-(B-1/2)r'}\, e^{(Ae^{-r'})}, \tag{7}$$

which is depicted in Fig.2 as a function of the same variables (but only for a quadrant, exploiting its circular symmetry). Observe that $\phi_o(r')$ is concentrated in a ring between $r' \sim 1$ and $r' \sim 3$, and for $r' \sim 0$ it is $\phi_o(r') \approx 0$. That means that the effective support of the pattern $\Phi^{tw}(r,\theta,t)$ is always $r > vt$, and the same comment is valid for $D(r,\theta,t)$ and $F(r,\theta,t)$. Also—having $G_o(r')$ and $g(r')$—we calculate

$$\alpha_o(r') = (A^2 - a^2)e^{-2r'} + (a + 2AB)e^{-r'} + \left[\left(B - \frac{1}{2}\right)^2 - 2ab\right] - be^{r'} - b^2 e^{2r'}.$$

A last ingredient is still needed in order to obtain the effective reaction term

$$F(r') = -f(r')\alpha_o(r')\phi_o(r'),$$

(namely $\left(\frac{\partial\phi_o}{\partial r'}\right)^{n-1}$) but it amounts to take the derivative of eq.(7) with respect to r'. It is worth noting that—being $\alpha_o(r')$ finite—the result of a zero in $\frac{\partial\phi_o}{\partial r'}$ will be only a harmless zero in $F(r')$. In Figures 3 and 4 we show $F(r')$ for $n = 1$ (reaction-diffusion case) and $n = 2$, respectively.

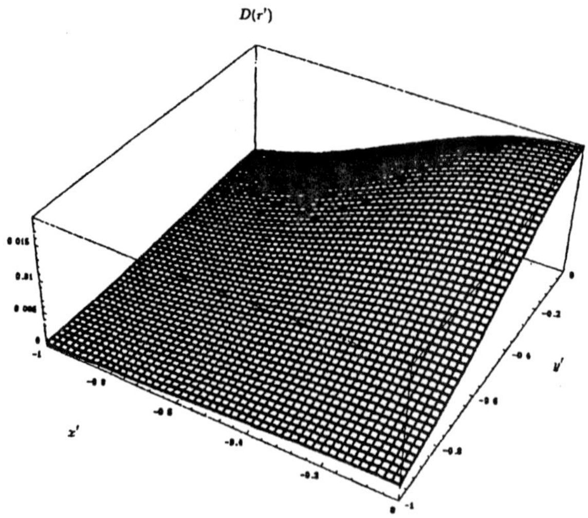

Figure 1. Dependence upon x' and y' of the diffusion term $D(r')$, for $n = 1$.

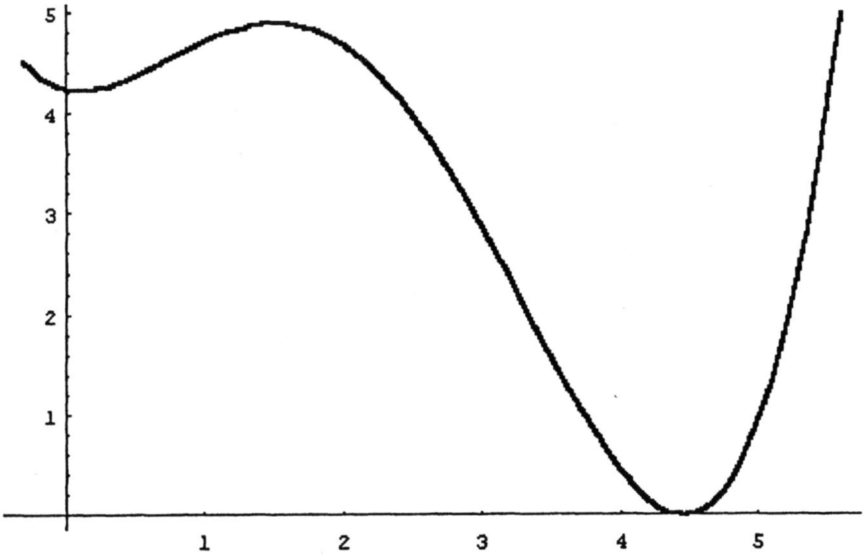

Figure 2. Dependence of $\Phi_o^{st}(r')$ upon x' and y'. In all the graphs we have taken $A = -50$, $B = 1$, $a = 1$ and $b = -1$.

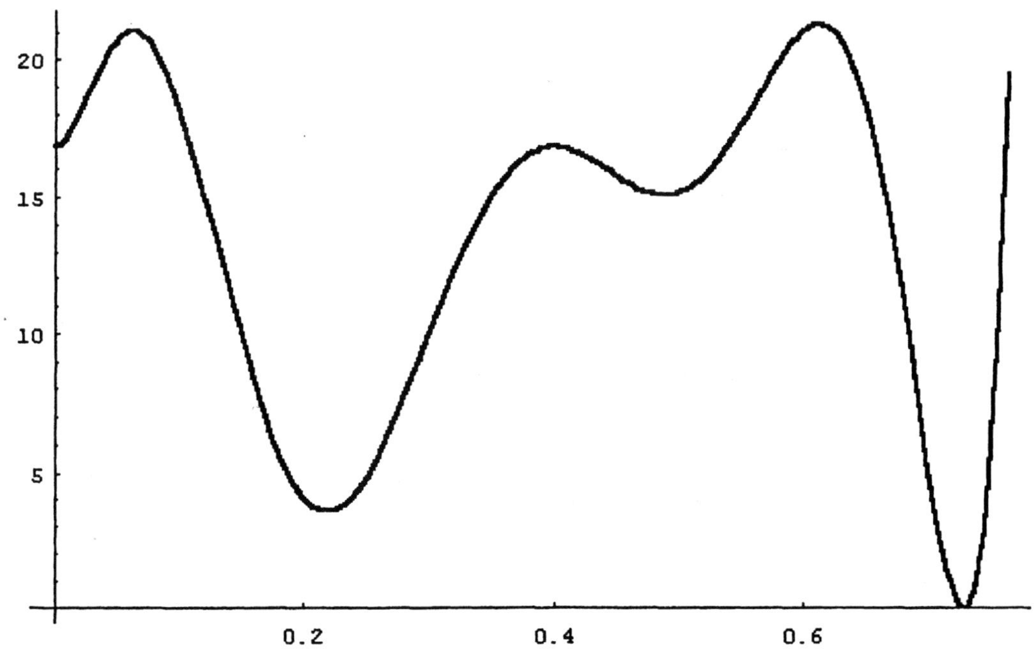

Figure 3. Dependence upon x' and y' of the reaction term $F(r')$, for $n = 1$ (reaction-diffusion system).

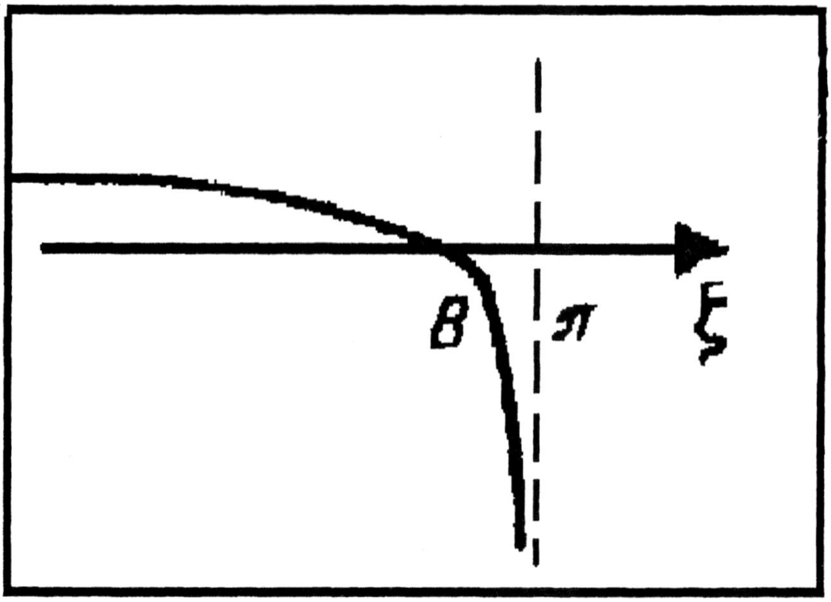

Figure 4. Dependence upon x' and y' of the reaction term $F(r')$, for $n = 2$. $F(r')$ acquires similar features to those of $\Phi_o^{st}(r')$.

As a general trend we see that—whereas for $n \leq 2$, $F(r')$ is a monotonic function—for $n \geq 2$ it develops a ring which coincides with that of $\phi_o(r')$. The amplitude of this ring decreases as n increases above 1, due to the fact that in our example it is everywhere $\left|\frac{d\phi_o(r')}{dr'}\right| < 1$ (recall that $F(r') \propto \left(\frac{d\phi_o(r')}{dr'}\right)^{n-1}$). Recalling also that the reaction term acts as an energy source for the system, we might interpret this fact by thinking that—in this case—the system becomes more efficient as n increases above 1, and less energy is needed for the wave $\phi_o(r')$ not to fade out. The opposite would occur where $\left|\frac{d\phi_o(r')}{dr'}\right| > 1$.

In principle (by following the steps in Ref. [8]) the method can be extended to more complicated physical situations—with a breakdown of angular symmetry, or where the pattern's motions are angular as well as radial. These would require in turn more complicated spatio-temporal dependences of $F(\vec{x}', t)$ and $D(\vec{x}', t)$—the reaction and diffusion terms, respectively. Nonetheless, we hope to have already convinced the reader of the potential usefulness of the semi-inverse method we are now introducing, for many applications in physics and related sciences. The possibility to generate solutions with desired qualitative features for a comprehensive class of nonlinear equations could be a valuable auxiliary tool in modelling, as well as a good starting point for other methods.

M. Bellini acknowledges Conacyt and CIC of Universidad Michoacana for financial support in the form of a research grant.

References

[1] M. C. Cross and P. C. Hohenberg, Rev. Mod. Phys. **65**, 851 (1993).
[2] E. Meron, Phys. Rep. **218**, 1 (1992).
[3] Y. Kuramoto, *Chemical Oscillations, Waves and Turbulence* (Springer, 1984); R. J. Field and M. Burger, eds., *Oscillation and Travelling Waves in Chemical Systems* (Wiley, 1985); Ya. B. Zeldovich et al., *Mathematical Theory of Combustion and Explosion* (Consultants Bureau, NY, 1985).
[4] J. D. Murray, *Mathematical Biology* (Springer, 1989).
[5] R. R. Nelson and S. G. Winter, *An Evolutionary Theory of Economic Change* (Harvard University Press, 1982); B. Silverberg, in *Technical Change and Economic Theory*, eds. G. Dosi et al. (Pinter, NY, 1988) p.531.
[6] M. Bellini, N. Giovambattista and R. Deza, Nuovo Cim. B **112**, 767 (1997).
[7] M. Bellini, R. Deza and N. Giovambattista, Phys. Lett. A **232**, 200 (1997).
[8] N. Giovambattista, M. Bellini and R. Deza, Inverse Problems, **16**, 811 (2000).
[9] Mauricio Bellini and Roberto R. Deza, Nuovo Cim. **B113**, 1231 (1998).
[10] J. Gratton and C. Vigo, *Evolution of self-similarity, and other properties of waiting-time solutions of the porous media equation: the case of viscous gravity currents*, Universidad de Buenos Aires preprint.
[11] A-L. Barabási and H. E. Stanley, *Fractal Concepts in Surface Growth* (Cambridge U. Press, 1996).

Theoretical Physics 2002, Part 1
Edited by Thomas F. George and Henk F. Arnoldus, pp. 83-100
ISBN 1-59033-435-3

Optical Potential in the Nuclear Matter Approximation with Generalized Skyrme Interaction

S. M. Kravchenko*and A. P. Soznik†

Abstract

The nucleon-nucleus optical potential is obtained in the nuclear matter approximation for the generalized density-dependent two-body Skyrme forces with allowance for the tensor nucleon-nucleon effective interaction. The rearrangement potential effect on the radial distributions of the optical potential is investigated. It is shown that two- and three-body effective interactions are not equivalent in calculating the imaginary part of the optical potential. Tensor Skyrme forces appear to have a marked influence on the imaginary part of the optical potential.

1 Introduction

It is known [1] that the microscopic nucleon-nucleus optical potential (OP) can be identified with the mass operator of the one-body Green function. According to this approach the OP in the nuclear matter approximation was calculated [2] on the basis of the Hartree-Fock theory with effective Skyrme interaction. The results obtained in [2] at zero nucleus temperature were then generalized [3–5] for finite temperatures.

It is natural that the expressions for the OP at zero and non-zero temperatures differ from one another. However, there are other differences in principle between the OP in [2] and [3–5] which can not be explained by the temperature dependence.

In this paper, we have shown that there are two reasons for divergence between the results of [2] and [3–5]. First of all, it is known from the investigations of even-even nuclei and nuclear matter by the Hartree-Fock method with effective interaction [6, 7] that three-body Skyrme forces (used in [2]) are equivalent to two-body forces linearly depending on density which are the particular case of the forces used in [3–5]. Since the real part of the OP in [2–5] coincides with the one-body Hartree-Fock potential, then, as shown below, the difference between the expressions for this

*Applied Physics Institute, National Academy of Sciences of Ukraine, Petropavlovskaya St. 58, Sumy, 40011, Ukraine, E-mail:iapuas@iap.infonet.com.ua

†V.N.Karazin Kharkov National University, Svobody Sq. 4, Kharkov, 61077, Ukraine, E-mail: soznik@mingu.kharkov.ua

part of the OP results from the fact that the authors of [3–5] have neglected the rearrangement (or saturation) potential which gives a substantial contribution to the optical potential.

We have shown that unlike the real part of the OP, the analytical expression for its imaginary part depends essentially on the type of forces used (two- or three-body). It means that two- and three-body Skyrme forces are not equivalent for obtaining the imaginary part of the OP. The expression for the imaginary part of the OP obtained with two-body Skyrme forces coincides with the corresponding expression in [3–5] at zero temperature where the rearrangement potential was not taken into account. However, the rearrangement potential defines the intensity and radial dependence of the real part of the OP largely, changing the dispersion law for the nucleon motion in nuclear matter. This, in turn, gives rise to the change of the intensity and shape of the imaginary part of the OP.

It is known [8, 9] that the effective Skyrme interaction contains the components describing the tensor interaction between the nucleons inside a nucleus, but these components are both neglected in [2] and in [3–5], as well as in other papers dealing with the various effective nucleon-nucleon interactions. In many cases, such simplification is justified since any density-dependent effective nucleon-nucleon interaction, including that of Skyrme, to a certain degree accounts for the contribution of tensor forces [10]. The tensor Skyrme forces were shown to have only weak influence on the one-particle spectra of even-even nuclei [8]. But, these forces were found to have strong influence on the central and tensor spin-spin potentials of nucleon interaction with odd nuclei [11].

Therefore, in this paper, we have included the tensor terms of the Skyrme interaction in the calculations of the optical potential. In Sec. 2, the Hartree-Fock equations with the effective Skyrme forces are discussed and the necessity of taking into account the rearrangement potential is shown. The real part of the OP is calculated in Sec. 3 while its imaginary part is obtained in Sec. 4. In Sec. 5, the numerical calculations of the OP obtained are presented for the neutron scattering by ^{40}Ca nuclei, as well as some aspects of tensor forces parameterization are discussed. Section 6 presents final conclusions.

2 Rearrangement Potential

Let us choose the effective nucleon-nucleon interaction in the form of the density-dependent generalized Skyrme forces [12] with tensor terms [8]

$$v \equiv v(i,j) = v_1 + v_2(\rho) + v_{te} + v_{to}, \tag{1}$$

$$v_1 = t_0(1 + x_0 P_\sigma)\delta(\mathbf{r}) + \frac{1}{2}t_1(1 + x_1 P_\sigma)\left[\mathbf{k}'^2\delta(\mathbf{r}) + \delta(\mathbf{r})\mathbf{k}^2\right] +$$
$$+ t_2(1 + x_2 P_\sigma)\mathbf{k}'\delta(\mathbf{r})\mathbf{k} + iW_0\left[\mathbf{k}' \times \delta(\mathbf{r})\mathbf{k}\right](\boldsymbol{\sigma}_i + \boldsymbol{\sigma}_j), \tag{2}$$

$$v_2(\rho) = \frac{1}{6}t_3(1 + x_3 P_\sigma)\rho^\gamma(\mathbf{R})\delta(\mathbf{r}) +$$
$$+\frac{1}{2}t_4(1 + x_4 P_\sigma)\left[\mathbf{k}'^2\rho(\mathbf{R})\delta(\mathbf{r}) + \delta(\mathbf{r})\rho(\mathbf{R})\mathbf{k}^2\right] +$$
$$+t_5(1 + x_5 P_\sigma)\mathbf{k}'\rho(\mathbf{R})\delta(\mathbf{r})\mathbf{k}, \tag{3}$$

$$v_{te} = \frac{1}{2}T\left\{\left[(\boldsymbol{\sigma}_i \cdot \mathbf{k}')(\boldsymbol{\sigma}_j \cdot \mathbf{k}') - \frac{1}{3}(\boldsymbol{\sigma}_i \cdot \boldsymbol{\sigma}_j)\mathbf{k}'^2\right]\delta(\mathbf{r}) +\right.$$
$$\left. +\delta(\mathbf{r})\left[(\boldsymbol{\sigma}_i \cdot \mathbf{k})(\boldsymbol{\sigma}_j \cdot \mathbf{k}) - \frac{1}{3}(\boldsymbol{\sigma}_i \cdot \boldsymbol{\sigma}_j)\mathbf{k}^2\right]\right\}, \tag{4}$$

$$v_{to} = U\left\{(\boldsymbol{\sigma}_i \cdot \mathbf{k}')\delta(\mathbf{r})(\boldsymbol{\sigma}_j \cdot \mathbf{k}) - \frac{1}{3}(\boldsymbol{\sigma}_i \cdot \boldsymbol{\sigma}_j)(\mathbf{k}' \cdot \delta(\mathbf{r})\mathbf{k})\right\}, \tag{5}$$

where v_1 is density-independent, $v_2(\rho)$ is density-dependent, v_{te} is even tensor, and v_{to} is odd tensor forces; $\mathbf{r} = \mathbf{r}_i - \mathbf{r}_j$, $\mathbf{R} = (\mathbf{r}_i + \mathbf{r}_j)/2$, and other notations are the same as in [3–6, 8, 12]. As mentioned above, the following version of Skyrme forces were used in [2]

$$v = v_1, \quad w \equiv w(i, j, k) = t_3\delta(\mathbf{r}_i - \mathbf{r}_j)\delta(\mathbf{r}_j - \mathbf{r}_k), \tag{6}$$

where w is the three-body interaction.

Let the interaction be defined according to equations (2), (6). Then one-body Hamiltonian can be written in the symbolic form

$$h = t\rho + v_1\rho + \frac{1}{2}w\rho\rho, \tag{7}$$

and the mean field U_{HF} takes the ordinary Hartree-Fock form

$$U_{HF} = \tilde{v}\rho, \tag{8}$$

$$\tilde{v} = v_1 + \frac{1}{2}w\rho. \tag{9}$$

When choosing the forces (1)–(3) the Hamiltonian h and potential $\tilde{v}$ are expressed as

$$h = t\rho + v_1\rho + v_2(\rho)\rho + \frac{1}{2}\frac{\partial v_2(\rho)}{\partial \rho}\rho\rho, \tag{10}$$

$$\tilde{v} = v_1 + v_2(\rho) + \frac{1}{2}\frac{\partial v_2(\rho)}{\partial \rho}\rho \equiv v + \frac{1}{2}\frac{\partial v}{\partial \rho}\rho. \tag{11}$$

We have omitted in (10), (11) the tensor forces v_{te}, v_{to}, because they are not essential for the discussion of the rearrangement potential. Substituting equations (1)–(3) with $t_4 = t_5 = 0$ and $x_3 = \gamma = 1$ into equation (11), we obtain the expressions for $\tilde{v}$ and U_{HF} which coincide with equations (8), (9). The statement [6] about the equivalency of the interactions (1)–(3) and (2), (6) for the description of nuclear matter and even-even nuclei properties is namely based on this fact. However, this equivalency does not signify that the mean field U_{HF} in the case of the interaction (1)–(3) can be obtained by the direct substitution of equation (1) into equation (6). The consistent mean field calculation for density-dependent effective interaction should include the potential in the form (11) with modified initial interaction v containing the term proportional to $\frac{\partial v}{\partial \rho}$.

The calculations of the real and imaginary parts of the OP in [3–5] violate the procedure for obtaining the mean field described by equations (8) and (11) since the authors of [3–5] have neglected the terms proportional to $\frac{\partial v}{\partial \rho}$. The terms of the mean field which are proportional to $\frac{\partial v}{\partial \rho}$ are referred to as rearrangement potential [13] or saturation potential [14, 15]. These terms are important for the self-consistency of the Hartree-Fock theory [7, 13], for obtaining the necessary nuclear saturation at the valid density, and for providing the valid density distribution [7, 13–15]. In particular, neglecting the rearrangement potential results in too high density in the center of nucleus and this, it will be shown below, in its turn, leads to too deep real part of the OP. It should be emphasized that the terms in the potential U_{HF}, proportional to $\frac{\partial v}{\partial \rho}$, have been obtained [14, 15] both by varying the energy and from Goldstone diagrams, and it was shown that both methods lead under certain conditions, to the same results.

3 Real Part of the Optical Potential

According to [1], in the first order of the perturbation theory, mass operator presents the real part of the OP and appears to be the mean Hartree-Fock potential. Taking into account equations (8) and (11), we obtain

$$U_{\lambda\lambda} \equiv (U_{HF})_{\lambda\lambda} = U_{\lambda\lambda}^{(0)} + U_{\lambda\lambda}^{(R)}, \tag{12}$$

$$U_{\lambda\lambda}^{(0)} = \sum_{\mu} \left[\langle \lambda\mu | v | \lambda\mu \rangle - \langle \lambda\mu | v | \mu\lambda \rangle \right], \tag{13}$$

$$U_{\lambda\lambda}^{(R)} = \frac{1}{2} \sum_{\mu\nu} \left[\langle \tilde{\lambda}\mu\nu \left| \frac{\partial v}{\partial \rho} \right| \tilde{\lambda}\mu\nu \rangle - \langle \tilde{\lambda}\mu\nu \left| \frac{\partial v}{\partial \rho} \right| \tilde{\lambda}\nu\mu \rangle \right], \tag{14}$$

where $U_{\lambda\lambda}^{(R)}$ is the rearrangement potential. The sign tilde for the state λ means that $|\tilde{\lambda}\rangle \equiv \varphi_\lambda(\mathbf{R})$, $\mathbf{R} = (\mathbf{r}_\mu + \mathbf{r}_\nu)/2$ [16].

As in [3–5], let us consider the unpolarized nuclear matter approximation and define the nuclear wave function $\varphi_\gamma(\mathbf{r})$ in the state γ as the plane wave normalized by volume. Then, taking into account equations (1)–(5), we obtain the expression for $U_{\lambda\lambda}$ which can be presented as follows

$$U_{\lambda\lambda}^{(0)} = Z_\lambda + K_\lambda^2 \frac{\hbar^2}{2m_\lambda}\left(\frac{m_\lambda}{m_\lambda^*} - 1\right), \tag{15}$$

$$\begin{aligned} U_{\lambda\lambda}^{(R)} &= \frac{1}{4}(g_4 + g_5)\rho\kappa - \frac{1}{4}(h_4 - h_5)\sum_q \rho_q \kappa_q \\ &+ \frac{1}{12}\gamma\rho^{\gamma-1}\left(g_3\rho^2 - h_3\sum_q \rho_q^2\right), \end{aligned} \tag{16}$$

$$\begin{aligned} Z_\lambda &= g_0\rho - h_0\rho_\lambda + \frac{1}{4}[g_1 + g_2 + (g_4 + g_5)\rho]\kappa \\ &- \frac{1}{4}[h_1 - h_2 + (h_4 - h_5)\rho]\kappa_\lambda + \frac{1}{6}(g_3\rho - h_3\rho_\lambda)\rho^\gamma, \end{aligned} \tag{17}$$

$$\frac{m_\lambda}{m_\lambda^*} = 1 + \frac{1}{4}\frac{2m_\lambda}{\hbar^2}\{[g_1 + g_2 + (g_4 + g_5)\rho]\rho - [h_1 - h_2 + (h_4 - h_5)\rho]\rho_\lambda\}, \tag{18}$$

$$g_i = t_i\left(1 + \frac{1}{2}x_i\right), \quad h_i = t_i\left(\frac{1}{2} + x_i\right), \quad i = 0, 1, 2, 3, 4, 5,$$

$$\kappa = \frac{3}{5}\left(\frac{3\pi^2}{2}\rho\right)^{2/3}\rho, \quad \kappa_q = \frac{3}{5}(3\pi^2\rho_q)^{2/3}\rho_q, \quad q = p, n,$$

where m_λ is the mass and K_λ is the wave vector of nucleon which is defined in the nuclear matter as in [2–5]

$$K_\lambda^2 = \frac{2m_\lambda}{\hbar^2}\left[\frac{M}{M + m_\lambda}E - U_{\lambda\lambda}\right]. \tag{19}$$

In equation (19), M is the mass of nucleus and E is the energy of incident nucleon in the laboratory frame. By excluding K_λ^2 from equations (12), (15), (16), and (19) and by applying the local density approximation [17, 18] we arrive at the expression for the real part of the OP

$$U_{\lambda\lambda}(r, E) = \frac{m_\lambda^*}{m_\lambda}\left(Z_\lambda + U_{\lambda\lambda}^{(R)}\right) + \left(1 - \frac{m_\lambda^*}{m_\lambda}\right)\frac{M}{M + m_\lambda}E, \tag{20}$$

where m_λ^*, Z_λ and $U_{\lambda\lambda}^{(R)}$ are defined by equations (16)–(18) where $\rho = \rho(r)$ and $\rho_q = \rho_q(r)$. Thus, the OP at the point r of a finite nucleus has the same value as in the nuclear matter with the corresponding densities ρ and ρ_q. It should be noted that the generalization of equations (15)–(20) for the case of non-zero temperatures can be directly derived by the method developed in [3–5].

The OP obtained above and the corresponding expressions in [3] (equation (11)) and in [5] (equation (5)) differ in all terms proportional to t_3, t_4, and t_5 which are just responsible for the density dependence of the effective interaction (1)–(5). In particular, for the symmetrical nuclear matter ($\rho_p = \rho_n = \rho/2$ and $\kappa_p = \kappa_n = \kappa/2$) we obtain the terms proportional to the factors t_4 and t_5 in equation (20) to be 4 times greater and the terms proportional to t_3 (for the case $\gamma = 1$ and $x_3 = 1$) to be 1.5 times greater than the corresponding terms in [3, 5].

At the same time one can obtain the real part of the OP from [2] (equation (26)) as the particular case of equation (20) directly for the Skyrme forces with the parameters $x_1 = x_2 = t_4 = t_5 = 0$, and $x_3 = \gamma = 1$ (the forces Sk2–Sk6). Because of using the unpolarized nuclear matter approximation, the tensor forces (4), (5) do not influence on the real part of the OP determined by the anti symmetrized matrix elements (13).

4 Imaginary Part of the Optical Potential

To calculate the imaginary part of the OP we use the scattering matrix expansion in a series of the perturbation theory over the residual interaction and separate the compact diagrams which define the mass operator from the self-energy part of this expansion which describes the particle or hole propagation processes [19]. The initial Hamiltonian with the interaction (1)–(5) can be presented as follows

$$H = h + h', \tag{21}$$

where the one-body self-consistent Hamiltonian h in the Hartree-Fock approximation is defined by equations (10), (11). The Hamiltonian of the residual (correlation) interaction h' can be written in the form [16, 19]

$$\begin{aligned} h' &= \frac{1}{2}\int d^3r d^3r' N[\Psi^+(\mathbf{r})\Psi^+(\mathbf{r}')V(\mathbf{r},\mathbf{r}',\rho)\Psi(\mathbf{r}')\Psi(\mathbf{r})] \\ &\quad -\frac{1}{2}\int d^3r N[\Psi^+(\mathbf{r})V_R(\mathbf{r})\Psi(\mathbf{r})], \end{aligned} \tag{22}$$

$$V(\mathbf{r},\mathbf{r}',\rho) = \langle \mathbf{r}'|v|\mathbf{r}\rangle = \langle \mathbf{r}'|v_1 + v_2(\rho)|\mathbf{r}\rangle, \tag{23}$$

$$V_R(\mathbf{r}) = \int d^3r_1 d^3r_2 \delta(\mathbf{r} - \mathbf{R})\frac{\partial V(\mathbf{r}_1,\mathbf{r}_2,\rho)}{\partial\rho}(1 - P_{12})\, R(\mathbf{r}_1,\mathbf{r})R(\mathbf{r}_2,\mathbf{r}), \tag{24}$$

$$R(\mathbf{r}_1,\mathbf{r}) = \langle \mathbf{r}_1|\rho|\mathbf{r}\rangle. \tag{25}$$

$\Psi^+(\mathbf{r})$ and $\Psi(\mathbf{r})$ in equation (22) are the creation and annihilation field operators, P_{12} is the exchange operator, symbol N means the normal product of the operators [19]. It follows from equation (22) that the residual interaction directly depends on the rearrangement potential $V_R(\mathbf{r})$.

The imaginary part of the mass operator (imaginary part of the OP) in the second order of the perturbation theory can be presented as [20]

$$\mathrm{Im}\, M_{\alpha\alpha}(\varepsilon) = -\pi \sum_{\lambda\mu\nu} \langle \alpha\nu|v|\lambda\mu\rangle\langle\lambda\mu|v(1-P)|\alpha\nu\rangle$$

$$\times (1-n_\lambda)(1-n_\mu)\, n_\nu \delta\left(\varepsilon + \varepsilon_\nu - \varepsilon_\lambda - \varepsilon_\mu\right), \tag{26}$$

where ε_μ is the one-particle energy, n_μ is the occupation number.

Substituting the effective interaction (1)–(5) into equation (26) one can obtain the following expression for the imaginary part of the OP

$$W_{\alpha\alpha}(r) = -\frac{1}{64\pi^5}\sum_{i=1}^{7} W_i, \tag{27}$$

$$W_1 = \left(2g_{00} + \frac{1}{18}g_{33}\rho^{2\gamma} + \frac{2}{3}g_{03}\rho^\gamma\right)\left[I_1(\tau_\alpha, n) + I_1(\tau_\alpha, p)\right]$$

$$-\left(2h_{00} + \frac{1}{18}h_{33}\rho^{2\gamma} + \frac{2}{3}h_{03}\rho^\gamma\right) I_1(\tau_\alpha, \tau_\alpha), \tag{28}$$

$$W_2 = \left(2g_{01} + 2g_{04}\rho + \frac{1}{3}g_{13}\rho^\gamma + \frac{1}{3}g_{34}\rho^{\gamma+1}\right)\left[I_2(\tau_\alpha, n) + I_2(\tau_\alpha, p)\right]$$

$$-\left(2h_{01} + 2h_{04}\rho + \frac{1}{3}h_{13}\rho^\gamma + \frac{1}{3}h_{34}\rho^{\gamma+1}\right) I_2(\tau_\alpha, \tau_\alpha), \tag{29}$$

$$W_3 = \left(\frac{1}{2}g_{11} + \frac{1}{2}g_{44}\rho^2 + g_{14}\rho + \frac{1}{3}T^2\right)\left[I_3(\tau_\alpha, n) + I_3(\tau_\alpha, p)\right]$$

$$-\left(\frac{1}{2}h_{11} + \frac{1}{2}h_{44}\rho^2 + h_{14}\rho + \frac{1}{3}T^2\right) I_3(\tau_\alpha, \tau_\alpha), \tag{30}$$

$$W_4 = 2\left(2g_{02} + 2g_{05}\rho + \frac{1}{3}g_{23}\rho^\gamma + \frac{1}{3}g_{35}\rho^{\gamma+1}\right)\left[I_4(\tau_\alpha, n) + I_4(\tau_\alpha, p)\right], \tag{31}$$

$$W_5 = 2\left(g_{12} + g_{15}\rho + g_{24}\rho + g_{45}\rho^2 + \frac{2}{3}UT\right)\left[I_5(\tau_\alpha, n) + I_5(\tau_\alpha, p)\right], \tag{32}$$

$$W_6 = 2\left(g_{22} + g_{55}\rho^2 + 2g_{25}\rho + \frac{2}{3}U^2\right)\left[I_6(\tau_\alpha, n) + I_6(\tau_\alpha, p)\right]$$

$$+2\left(h_{22} + h_{55}\rho^2 + 2h_{25}\rho + \frac{2}{3}U^2\right) I_6(\tau_\alpha, \tau_\alpha), \tag{33}$$

$$W_7 = (4W_0^2 + U^2 - T^2)\left[I_7(\tau_\alpha, n) + I_7(\tau_\alpha, p)\right]$$

$$+(4W_0^2 + U^2 + T^2) I_7(\tau_\alpha, \tau_\alpha), \tag{34}$$

where $\rho = \rho(r)$, $g_{ij} = t_i t_j \left(1 + x_i x_j + \frac{x_i + x_j}{2}\right)$,

$$h_{ij} = t_i t_j \left(x_i + x_j + \frac{1 + x_i x_j}{2}\right), \quad i, j = 0, 1, \ldots, 5;$$

$$I_i(\tau_\alpha, \tau_\mu) = \int \mathrm{d}\mathbf{K}_\mu \mathrm{d}\mathbf{K}_\lambda \mathrm{d}\mathbf{K}_\nu f_i \left(\mathbf{K}_\alpha, \mathbf{K}_\mu, \mathbf{K}_\lambda, \mathbf{K}_\nu\right) \delta(E + \varepsilon_\mu - \varepsilon_\lambda - \varepsilon_\nu)$$

$$\times \delta\left(\mathbf{K}_\alpha + \mathbf{K}_\mu - \mathbf{K}_\lambda - \mathbf{K}_\nu\right) n_\mu (1 - n_\lambda)(1 - n_\nu), \quad i = 1, \ldots, 7, \qquad (35)$$

$$f_1 = 1,\ f_2 = \mathbf{K}_{\alpha\mu}^2 + \mathbf{K}_{\lambda\nu}^2,\ f_3 = \left(\mathbf{K}_{\alpha\mu}^2 + \mathbf{K}_{\lambda\nu}^2\right)^2,\ f_4 = \mathbf{K}_{\alpha\mu} \cdot \mathbf{K}_{\lambda\nu},$$
$$f_5 = \left(\mathbf{K}_{\alpha\mu}^2 + \mathbf{K}_{\lambda\nu}^2\right)\left(\mathbf{K}_{\alpha\mu} \cdot \mathbf{K}_{\lambda\nu}\right),\ f_6 = \left(\mathbf{K}_{\alpha\mu} \cdot \mathbf{K}_{\lambda\nu}\right)^2,\ f_7 = \left(\mathbf{K}_{\alpha\mu} \times \mathbf{K}_{\lambda\nu}\right)^2,$$

$$\mathbf{K}_{\alpha\mu} = \frac{1}{2}\left(\mathbf{K}_\alpha - \mathbf{K}_\mu\right), \mathbf{K}_{\lambda\nu} = \frac{1}{2}\left(\mathbf{K}_\lambda - \mathbf{K}_\nu\right), \tau_\lambda = \tau_\alpha, \tau_\nu = \tau_\mu.$$

In equations (28)–(35), τ_μ equals to p for protons and n for neutrons. It should be noted that the integrals (35) have been analytically calculated in [2].

It is mentioned above that the rearrangement potential does not give direct contribution to the imaginary part of the OP in the second order of the perturbation theory over the residual interaction. Thus, equations (27)–(34) formally coincide with the corresponding expressions from [3] at zero temperature and $T = U = 0$ (equations (19)–(20)). The integrals I_1–I_7 are defined by the value of the incident nucleon wave vector K_α which, in its turn, depends on the real part of the OP. In fact, according to the dispersion law (19), the rearrangement potential gives the effective contribution to the imaginary part of the optical potential.

It follows from equations (30), (32)–(34), the tensor effective interaction (4), (5) has influence on the imaginary part of the OP, the corresponding terms being quadratic over the parameters T and U.

5 Analysis of the Optical Potential

5.1 Real part of the OP

The radial distributions of the real part of the OP for the neutron scattering by ^{40}Ca nuclei are shown in Fig. 1. The densities $\rho_n(r)$ and $\rho_p(r)$ are described, as in [2–5], by the empirical Negele's formula [18].

The dashed curves in Fig. 1 are obtained by neglecting the rearrangement potential, that is without the terms proportional to $\frac{\partial v}{\partial \rho}$. At the same time, the valid local density was used at each point r that just caused too strong an attraction. The solid curves in Fig. 1 include the rearrangement potential, and therefore, at the valid local density, the excess of attraction in the center of the nucleus is considerably reduced due to the saturation properties of this potential.

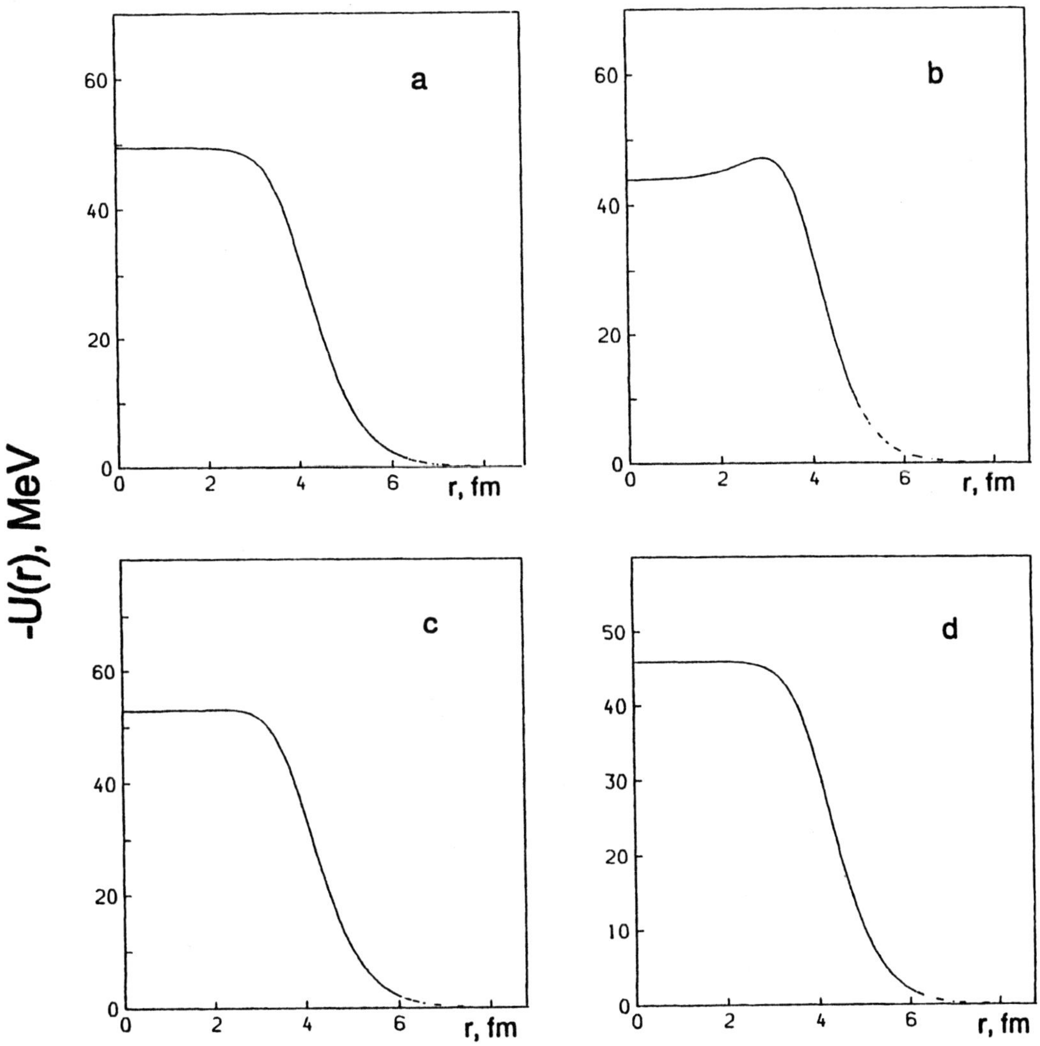

Figure 1: The real part of the optical potential for the $n+^{40}Ca$ elastic scattering at $E = 0$. Solid curves are the potentials (20), dashed curves are the potentials from [5] with the SkM* [21] (a), Sk3 [22] (b), SkT [23] (c), and Ska [24] (d) forces.

The differences between the potentials presented in Fig. 1 can be characterized by the value of the volume integral per nucleon which defines the cross section of nucleon scattering by nuclei

$$J_V = -\frac{1}{A}\int d^3r V(r), \tag{36}$$

One can see from Table 1 that neglecting the rearrangement potential $U^{(R)}_{\lambda\lambda}$ in equation (20) results in the values of J_V (the second line) which are considerably greater than those for the potential (20) (the first line). The experimental [25] value $J_V^{\exp} = 501$ MeV·fm^3 is close to the values of J_V for the potential (20) calculated with the SkM* and SkT forces and considerably greater than those for the Sk3 and Ska forces.

The second term in equation (20) defines the dependence of the real part of the OP on the incident nucleon energy. This dependence is accounted for by the value of the nucleon effective mass $m^*_\lambda(r)$ in the nucleus. For the nuclear matter the value $1 - m^*_\lambda/m_\lambda$ equals to 0.221, 0.266, 0.0, and 0.412 for the SkM*, Sk3, SkT, and Ska forces, respectively. The energy dependence of the experimental real part of the OP at $r = 0$ is defined by the term aE [25] where the constant $a = 0.254$ is close to the theoretical values for the SkM* and Sk3 forces.

Table 1. Volume integrals per nucleon J_V (MeV·fm^3) of the real part of the OP at $E = 0$ (n+^{40}Ca elastic scattering).

No.	SkM*	Sk3	SkT	Ska
1	499	463	515	465
2	590	533	632	516

5.2 Imaginary part of the OP

The results of the numerical calculations of the imaginary part of the OP are presented in Figs. 2–4 (without tensor effective interaction) and Fig. 5 (with including tensor interaction). As an example, the radial distributions are shown in Fig. 2, obtained with the use of two-body density-dependent S2 forces with ($W(r)$) and without ($W'(r)$) the rearrangement potential as well as with the use of the three-body S2 forces ($W''(r)$) at energies $E = 0$ and $E = 50$ MeV.

We see from Fig. 2 that the potentials $W(r)$, $W'(r)$, and $W''(r)$ differ from one another both by their intensities and shapes more considerably in the interior region and less considerably in the surface region of the nucleus. The difference between the potentials $W(r)$ and $W''(r)$ is accounted for by the non-equivalency of two- and three-body Skyrme forces for the calculation of the imaginary part of the OP while $W(r)$ and $W'(r)$ differ from one another due to including the rearrangement

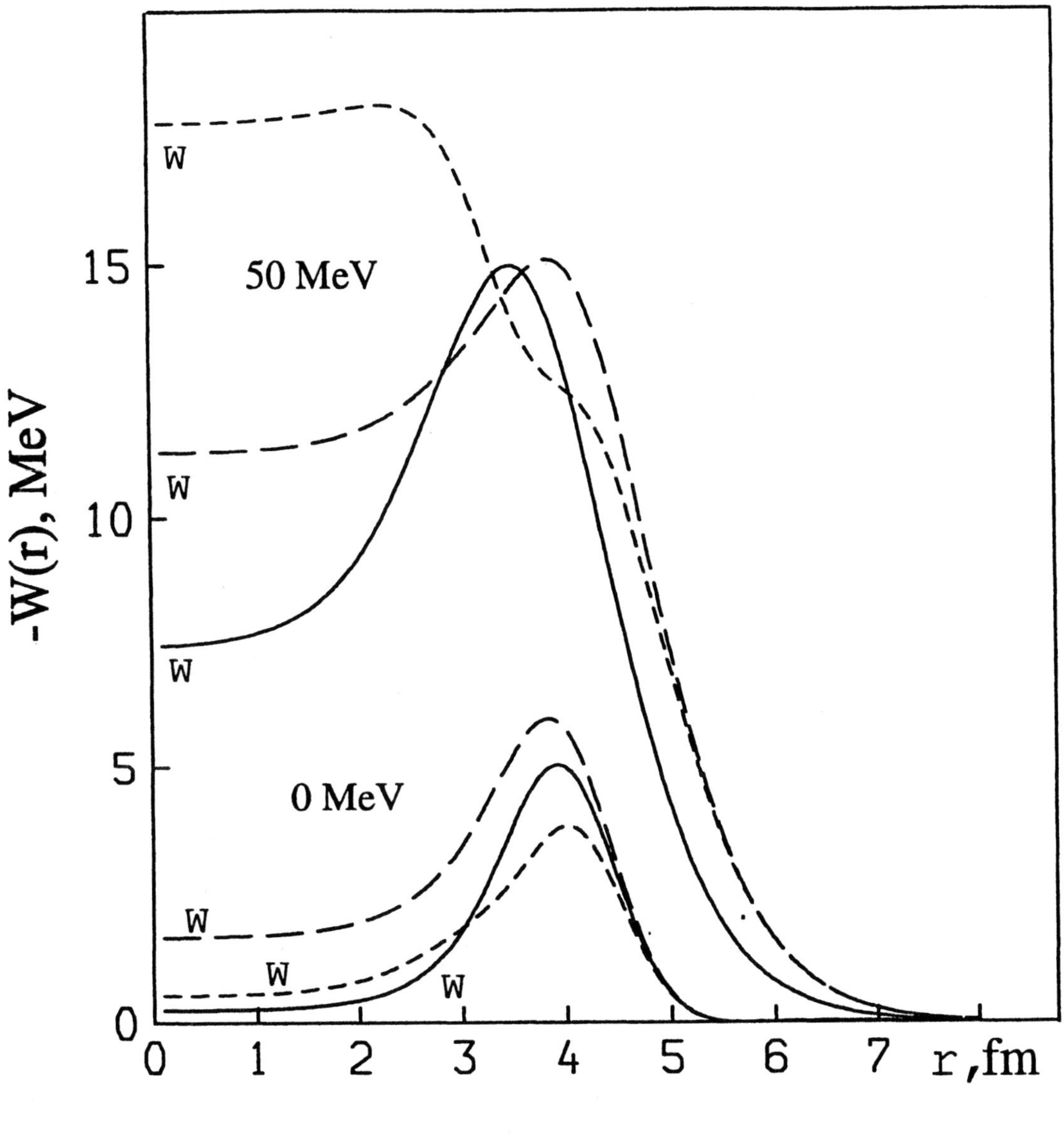

Figure 2: The radial distributions of the imaginary part of the OP for the S2 forces at $E = 0$ and $E = 50$ MeV (n+^{40}Ca elastic scattering).

potential in $W(r)$. It should be noted for the S2 forces that taking into account the rearrangement potential leads to the decrease of the volume absorption in comparison with the surface one.

The radial distributions $W(r)$ and $W'(r)$, calculated with the SkM* forces at $E = 50$ MeV, are shown in Fig. 3. Taking into account the rearrangement potential leads to the decrease of approximately 1.6 times the imaginary part of the OP in the center of the nucleus. At the same time, radial distributions are of volume character, unlike the S2 forces case.

The radial distributions $W(r)$, at various scattering energies for the SkM* forces, are presented in Fig. 4. It is shown that the surface character of the imaginary part of the OP changes with energy increasing to volume one. It should be noted that such behavior of $W(r)$, obtained for the SkM* forces, coincides with the energy dependence of the imaginary part of the OP, calculated with the three-body Skyrme forces [2]. Such energy dependence is also in good agreement with the experimentally obtained W dependence on energy: at small energies, the nucleon absorption by nuclei is largely of surface character while with the energy increase the volume absorption intensity increases, and at $E = 50$ MeV the potential W has solely a volume character.

Let us consider the influence of the tensor effective interaction (4), (5) on the radial distribution of the imaginary part of the OP. Two methods of tensor Skyrme forces parameterization have been suggested [8]. The first is the calculation of the parameters T and U on the basis of known free nucleon-nucleon scattering potential, while the second method is contained in fitting these parameters to describe the one-particle spectra of even-even nuclei. The second method appears to be preferred [11] in researching the elastic nucleon scattering by nuclei. According to [8, 11] for the SkM* forces one can obtain the following values of tensor interaction parameters: $T = -U = 163.5$ MeV·fm^5.

The radial distribution $W(r)$ with including the tensor Skyrme forces is presented in Fig. 5. One can see that taking into account the tensor forces leads to increasing the $W(r)$ depth in the central region of the nucleus and has no influence on its behavior on the nuclear surface.

In Fig. 5, partial contributions of the tensor interaction components corresponding to the even and odd nucleon relative motion states are also presented. The odd component contribution to the imaginary part of the OP (in the center of the nucleus) is several times greater than the even component. This is the result of anti symmetrization of the effective interaction matrix elements defining the imaginary part of the OP.

To characterize the imaginary part of the OP, we have also calculated the volume integrals per nucleon J_W (defined like integrals (36)) and root-mean-square radii

$$r_W^2 = \frac{\int d^3r r^2 W(r)}{\int d^3r W(r)}. \tag{37}$$

Table 2 shows that J_W and r_W values are considerably distinguished for various

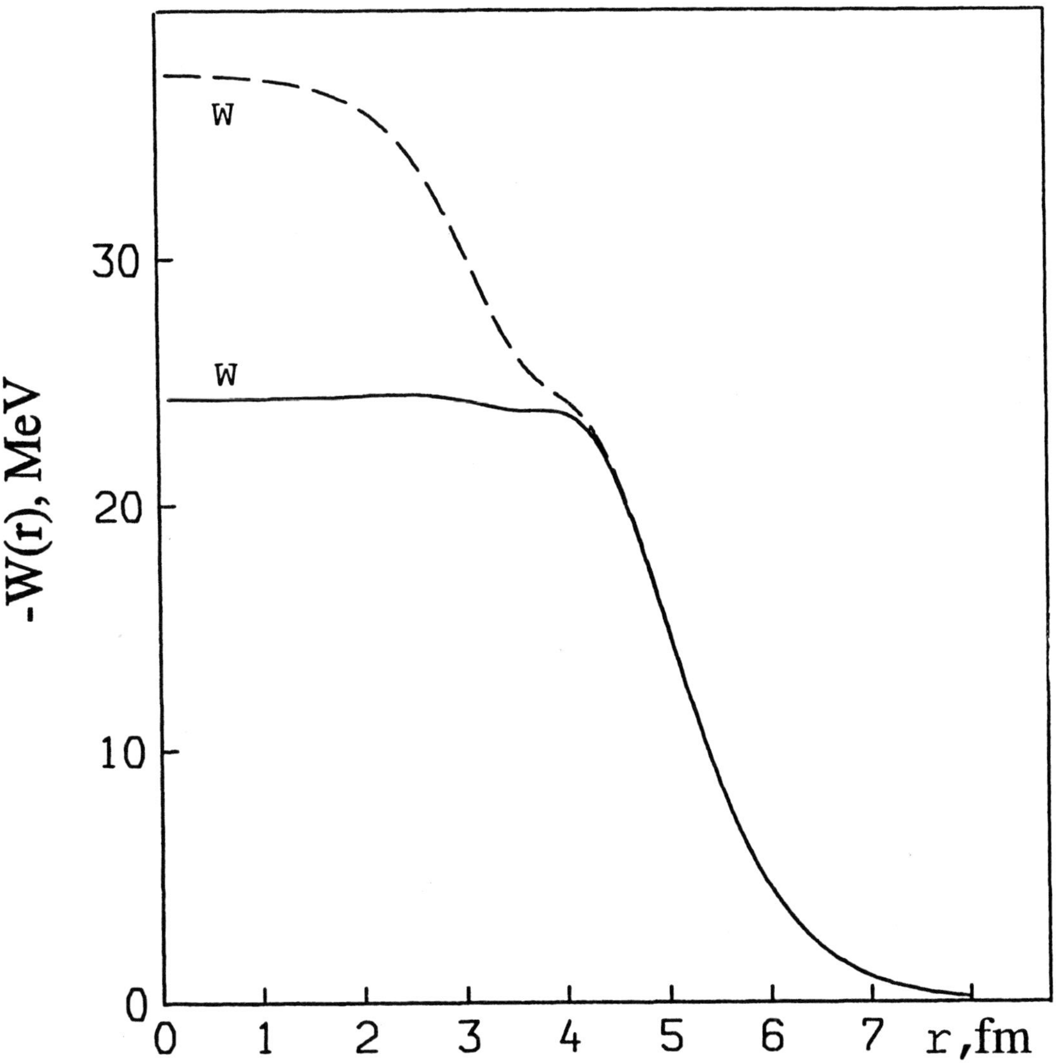

Figure 3: The same as in Fig. 2, for the SkM* forces at $E = 50$ MeV.

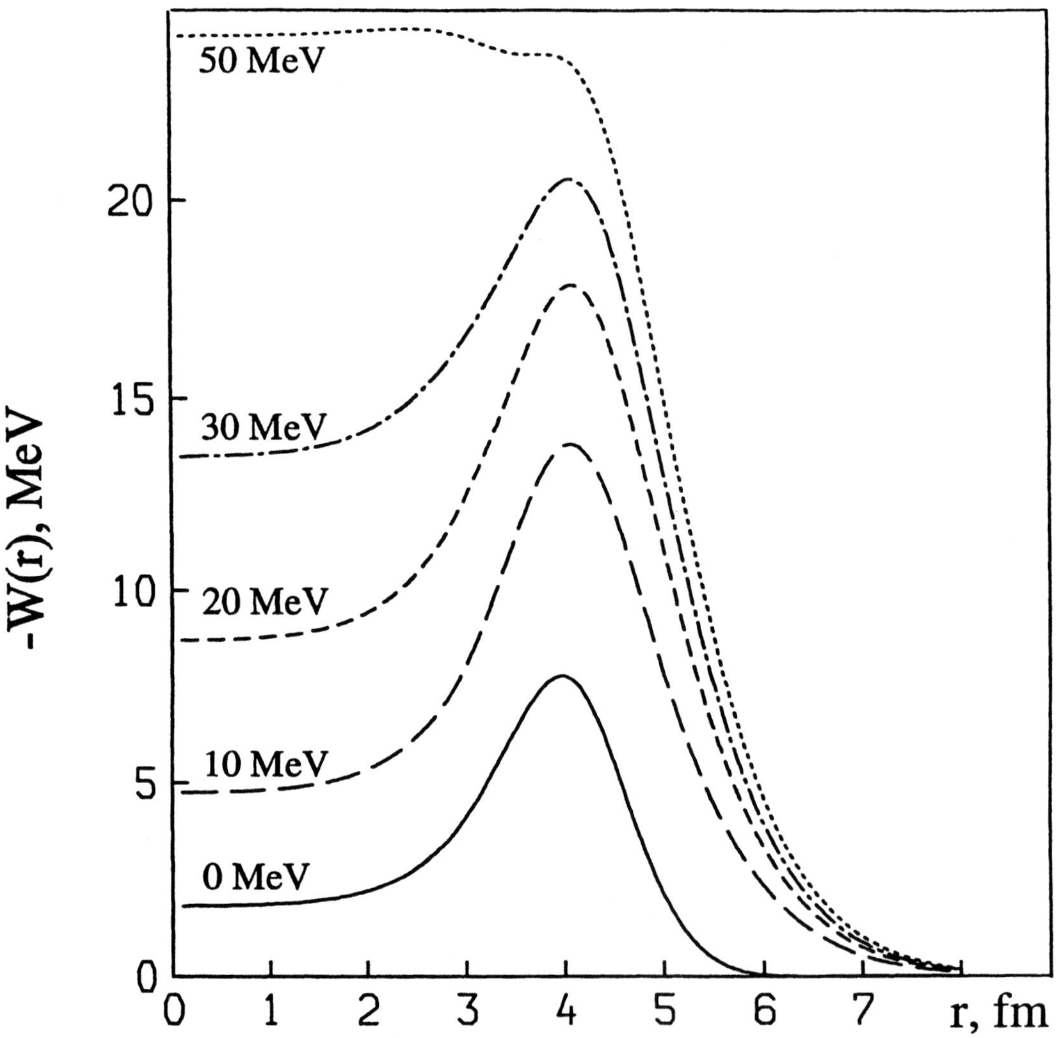

Figure 4: The energy dependence of the radial distribution $W(r)$ for the SkM* forces (n+^{40}Ca elastic scattering).

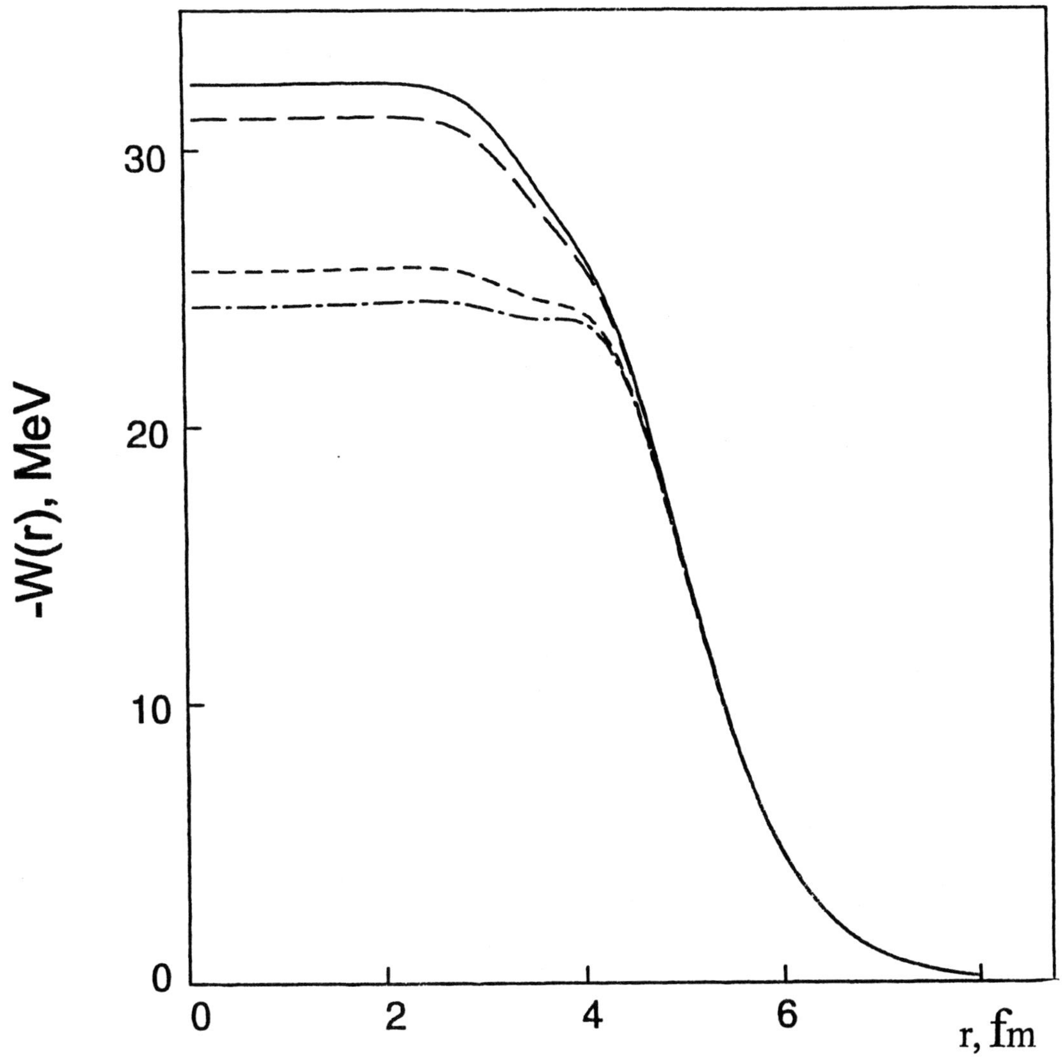

Figure 5: The radial distributions of the imaginary part of the OP (n+^{40}Ca elastic scattering) at E = 50 MeV for the SkM* forces with (solid curve) and without (dash-dotted curve) allowance for the tensor interaction, as well as including only the odd (T = 0, long dash curve) and only the even (U = 0, short dash curve) tensor components.

Skyrme forces, and the SkM* forces lead to the highest J_W values. It should also be noted that when taking into account the rearrangement potential, the results in the volume integral decrease and root-mean-square radii increase. On the other hand, when taking into account the tensor Skyrme forces, the inverse effect results.

Table 2. Volume integrals per nucleon J_W and root-mean-square radii r_W of the imaginary part of the OP (n+^{40}Ca elastic scattering). $\tilde{W}(r)$ is the potential with allowance for the tensor interaction.

Potential	Forces	J_W, MeV·fm^3		r_W, fm	
		$E = 0$	$E = 50$ MeV	$E = 0$	$E = 50$ MeV
	S2	35.25	201.71	4.05	4.40
$W(r)$	Ska	40.75	230.27	4.12	4.48
	SkM*	71.31	414.89	4.08	4.50
	S2	46.22	212.22	3.87	4.32
$W'(r)$	Ska	59.11	245.36	3.93	4.38
	SkM*	122.05	454.10	3.83	4.37
$W''(r)$	S2	30.09	217.14	4.03	4.19
$\tilde{W}(r)$	Ska	43.91	268.48	4.06	4.34
	SkM*	75.40	462.40	4.03	4.39

6 Conclusions

The OP calculations are performed in the nuclear matter approximation considering the rearrangement potential which arises due to the dependence of the two-body Skyrme forces on density, as well as with including the tensor effective forces.

The importance of taking into account the rearrangement potential is shown for investigations of both the real and imaginary parts of the OP. The real part of the OP reproduces the experimental data [25] more effectively when the SkM* forces are used. The shape and intensity of the imaginary part of the OP also greatly depend on the Skyrme forces used. The SkM* forces lead to the volume type of potential W at $E = 50$ MeV, which is in agreement with [25]. When using the S2 forces, the potential W at $E = 50$ MeV considerably conserves the surface absorption. At the same time, the absorption intensity (the value of W) is in good agreement with the experimental data [25] when the S2 forces are used, but poor for the SkM* forces. Also, it should be noted that the rearrangement potential reduces the absorption intensity considerably and brings the W value nearer to the experimental one [25].

Tensor effective Skyrme interaction does not influence the real part of the OP in the approximation being considered, but gives substantial contribution to its imaginary part. The most of this contribution is accounted for by the odd component of the tensor effective interaction.

7 Acknowledgment

The authors thank S. M. Shepeleva for assistance in preparing the English version of this paper.

References

1. J. S. Bell and E. J. Squires, A formal optical model, *Phys. Rev. Lett.* **3**, 96-97 (1959).

2. Q. Shen, J. Zhang, Y. Tian, Semi-microscopic optical potential calculation by the nuclear matter approach, *et al.*, *Z. Phys. A* **303**, 69-83 (1981).

3. L. Ge, Y. Zhuo and N. Wolfgang, Temperature-dependent optical potential and mean free path based on Skyrme interaction, *Nucl. Phys. A* **459**, 77-92 (1986).

4. Y. Zhuo, Y. Han and X. Wu, Further investigation on temperature-dependent optical potential and mean free path, *Prog. Theor. Phys.* **79**, 110-119 (1988).

5. G.-Q. Li, J.-Q. Shi and Q. Gao, The thermal self-consistent semiclassical approach and microscopic nucleon-nucleus potentials at finite temperature, *Nucl. Phys. A* **515**, 273-296 (1990).

6. D. Vautherin and D. M. Brink, Hartree-Fock calculations with Skyrme's interaction. I. Spherical nuclei, *Phys. Rev. C* **5**, 626-647 (1972).

7. H. A. Bethe, Theory of nuclear matter,*Ann. Rev. Nucl. Sci.* **21**, 93-217 (1971).

8. Fl. Stancu, D. M. Brink and H. Flocard, The tensor part of Skyrme's interaction, *Phys. Lett. B* **68**, 108-112 (1977).

9. K.-F. Liu, H. Luo, Zh. Ma, Skyrme-Landau parametrization of effective interaction. (I) Hartree-Fock ground states, *et al.*, *Nucl. Phys. A* **534**, 1-24 (1991).

10. H. A. Bethe, Thomas-Fermi theory of nuclei, *Phys. Rev.* **167**, 879-907 (1968).

11. S. M. Kravchenko and A. P. Soznik, Potential of nucleon interaction with odd nuclei in the Hartree-Fock theory, *Int. J. Mod. Phys. E* **8**, 137-157 (1999).

12. S. Krewald, V. Klemt, J. Speth and A. Faessler, On the use of Skyrme forces in self-consistent RPA calculations, *Nucl. Phys. A* **281**, 166-208 (1977).

13. K. A. Brueckner, *Theory of Nuclear Structure. The Many-Body Problem*, (Wiley, New York, 1959), p. 131.

14. B. H. Brandow, Compact-cluster expansion for the nuclear many-body problem, *Phys. Rev.* **152**, 863-882 (1966).

15. B. H. Brandow, Linked-cluster expansions for the nuclear many-body problem, *Rev. Mod. Phys.* **39**, 771-828 (1967).

16. K. Kumar, *Perturbation Theory and the Nuclear Many-Body Problem*, (North Holland, Amsterdam, 1962), p. 175.

17. K. A. Brueckner, J. L. Gammel and H. Weitzner, Theory of finite nuclei, *Phys. Rev.* **110**, 431-445 (1958).

18. J. W. Negele, Structure of finite nuclei in the local-density approximation, *Phys. Rev. C* **1**, 1260-1321 (1970).

19. D. A. Kirzhnits, *Field Theoretical Methods in Many-Body Systems*, (Pergamon, Oxford, 1967), p. 139.

20. S. M. Kravchenko, V. I. Kuprikov and A. P. Soznik, Semimicroscopic optical potential with effective density-dependent nucleon-nucleon interaction, *Int. J. Mod. Phys. E* **7**, 465-483 (1998).

21. J. Bartel, P. Quentin, M. Brack, Towards a better parametrization of Skyrme-like effective forces: a critical study of the SkM force, *et al.*, *Nucl. Phys. A* **386**, 79-100 (1982).

22. M. Beiner, H. Flocard, N. van Giai, P. Quentin, Nuclear ground-state properties and self-consistent calculations with Skyrme interaction (I) Spherical description, *Nucl. Phys. A* **238**, 29-69 (1975).

23. F. Tondeur, A Skyrme functional Hartree-Fock calculations of nuclear mass and density distribution, *Phys. Lett. B* **123**, 139-142 (1983).

24. H. S. Köhler, Skyrme force and the mass formula, *Nucl .Phys. A* **258**, 301-316 (1976).

25. G. M. Honoré, W. Tornow, C. R. Howell *et al.*, Coupled-channel analysis of nucleon scattering from ^{40}Ca up to 80 MeV, *Phys. Rev. C* **33**, 1129-1140 (1986).

Theoretical Physics 2002, Part 1
Edited by Thomas F. George and Henk F. Arnoldus, pp. 101-121

ISBN 1-59033-435-3

Non-Factorizable Contributions in B Decays Revisited

K. Terasaki*

Abstract

$\bar{B} \to D\pi$, $D^*\pi$, $J/\psi\bar{K}$ and $J/\psi\pi$ decays are studied by decomposing their amplitude into a sum of factorizable and non-factorizable ones. The former is estimated by using the naive factorization while the latter is calculated by using a hard pion approximation in the infinite momentum frame.

Although experiments provide only upper bounds of the branching ratios for the color suppressed $\bar{B}^0 \to D^0\pi^0$ and $D^{*0}\pi^0$ decays, consistency among the branching ratios for the $\bar{B} \to D\pi$ and $D^*\pi$ decays including the charged modes leads to phenomenologically allowed lower limits of $\mathcal{B}(\bar{B}^0 \to D^0\pi^0)$ and $\mathcal{B}(\bar{B}^0 \to D^{*0}\pi^0)$. Our result is compared with the phenomenologically estimated branching ratios as well as the measured ones. As the consequence, it is seen that the non-factorizable amplitude is rather small in the color favored $\bar{B} \to D\pi$ and $D^*\pi$ decays but can still efficiently interfere with the main (factorized) amplitude. In the color suppressed $\bar{B} \to J/\psi\bar{K}$ and $J/\psi\pi$ decays, non-factorizable contribution is more important. A sum of the factorized and non-factorizable amplitudes can improve the result from the factorization, although it is sensitive to model dependent values of form factors involved.

I. INTRODUCTION

The factorization (or vacuum insertion) prescription was first proposed long time ago [1] and, after the discovery of the standard model, it was revived [2], has been improved and applied extensively to hadronic weak decays of heavy

---

*Yukawa Institute for Theoretical Physics, Kyoto University, Kyoto 606-8502, Japan

mesons [3,4]. The improved one has been supported by two independent arguments, i.e., one is the large N_c (color degree of freedom) argument [5] that the factorizable amplitude which is given by the leading terms in the large N_c expansion dominates in hadronic weak decays and the other is the color transparency argument [6] that the factorization works well under a particular kinematical condition, i.e., a heavy quark decays into another heavy quark plus a pair of light quark and anti-quark which are emitted colinearly with sufficiently high energies, for example, as $b \to c + (\bar{u}d)_1$, where $(\bar{u}d)_1$ denotes a color singlet $(\bar{u}d)$ pair. In the former case, if the factorization works well in B decays, it will again work well in K and charm decays since the large N_c argument is independent of flavors. On the other hand, in the latter case, it cannot work well in K and charm decays where the kinematics cannot satisfy the above condition, even if it works well in B decays.

We here consider two body decays of charm mesons to check if the large N_c argument works well in hadronic weak interactions. A naive application of the factorization prescription to charm decays leads to the color suppression [suppression of color mismatched decays, $D^0 \to \bar{K}^0\pi^0$, $\bar{K}^{*0}\pi^0$, etc., described by $c \to (s\bar{d})_1 + u$]. It means, for example, that the amplitude for the D meson decays into isospin $I = \frac{1}{2}$ $(\bar{K}\pi)$ final states is approximately cancelled by the one into $I = \frac{3}{2}$ $(\bar{K}\pi)$ final states and hence the phases of these amplitudes are nearly equal to each other. Therefore the factorized amplitudes for two body decays of charm mesons should be approximately real except for the overall phase. However the measured rates for these decays are not compatible with the color suppression and the amplitudes for $D \to \bar{K}\pi$ and $\bar{K}^*\pi$ decays have large phase differences between the amplitudes for decays into the $I = \frac{1}{2}$ and $\frac{3}{2}$ final states [7]. To get rid of this problem, the factorization has been implemented by taking account for final state interactions. However, amplitudes with final state interactions are given by non-leading terms in the large N_c expansion as will be discussed later, so that the large N_c argument does not work well in charm decays and hence also in B decays since the large N_c argument is independent of flavors. On the other hand, if the color transparency argument works well, the factorization will be a good approximation in the $\bar{B} \to D\pi$ and $D^*\pi$ decays while, in the $\bar{B} \to J/\psi\bar{K}$ and $J/\psi\pi$, the factorization cannot be a good approximation (in contrast with in the large N_c argument) but non-factorizable long distance contribution can play a role.

In this article, we study $\bar{B} \to D\pi$, $D^*\pi$, $J/\psi\bar{K}$ and $J/\psi\pi$ decays. In the next section, we will review briefly the effective weak Hamiltonian and our basic perspective. In Sec. III, the $\bar{B} \to D\pi$ and $D^*\pi$ decays will be studied by decomposing their amplitude into a sum of factorizable and non-factorizable ones. The former will be estimated by using the naive factorization while the latter is calculated by using a hard pion approximation in the infinite momentum frame (IMF). We will estimate phenomenologically allowed branching ratios from the measured ones and compare our result with the estimated ones as well as the measured ones. As the consequence, it will be seen that the factorization works fairly well but non-factorizable contributions are still not negligible. In Sec. IV, the color suppressed decays, $\bar{B} \to J/\psi \bar{K}$ and $J/\psi\pi$, will be investigated in the same way. A brief summary will be given in the final section.

II. EFFECTIVE WEAK HAMILTONIAN

Before we study amplitudes for B decays, we review briefly the $|\Delta B| = 1$ effective weak Hamiltonian. Its main part is usually written in the form

$$H_w = \frac{G_F}{\sqrt{2}} V_{ud} V_{bc} \Big\{ c_1 O_1 + c_2 O_2 \Big\} + h.c., \tag{1}$$

where c_1 and c_2 are the Wilson coefficients of the four quark operators, O_1 and O_2, respectively, given by products of color singlet left-handed currents,

$$O_1 =: (\bar{d}u)_{V-A}(\bar{c}b)_{V-A} : \quad \text{and} \quad O_2 =: (\bar{c}u)_{V-A}(\bar{d}b)_{V-A} : . \tag{2}$$

The renormalization scale (μ) dependence of c_i and O_i is not explicitly described unless it is required. V_{ij}'s denote the CKM matrix elements [8] which are taken to be real since CP invariance is always assumed in this article.

When we calculate factorizable amplitudes for the $\bar{B} \to D\pi$ and $D^*\pi$ decays later, we use, as usual, the so-called BSW Hamiltonian [3,4]

$$H_w^{\mathrm{BSW}} = \frac{G_F}{\sqrt{2}} V_{ud} V_{cb} \Big\{ a_1 O_1 + a_2 O_2 \Big\} + h.c. \tag{3}$$

which can be obtained from Eq.(1) by using the Fierz reordering, where the operators O_1 and O_2 in Eq.(3) should be no longer Fierz reordered. The coefficients a_1 and a_2 are given by

$$a_1 = c_1 + \frac{c_2}{N_c}, \quad a_2 = c_2 + \frac{c_1}{N_c}, \tag{4}$$

where N_c is the color degree of freedom. When H_w^{BSW} is obtained, an extra term which is given by a color singlet sum of products of colored currents,

$$\tilde{H}_w = \frac{G_F}{\sqrt{2}} V_{ud} V_{cb} \Big\{ c_2 \tilde{O}_1 + c_1 \tilde{O}_2 \Big\} + h.c., \tag{5}$$

comes out, where

$$\tilde{O}_1 = 2\sum_a : (\bar{d}t^a u)_{V-A}(\bar{c}t^a b)_{V-A} : \quad \text{and} \quad \tilde{O}_2 = 2\sum_a : (\bar{c}t^a u)_{V-A}(\bar{d}t^a b)_{V-A} : \tag{6}$$

with the generators t^a of the color $SU_c(N_c)$ symmetry.

To realize physical amplitudes for B decays by matrix elements of $\tilde{H}_w$, soft gluon(s) should be exchanged between quarks which belong to different meson states or different currents. Therefore, amplitudes given by $\tilde{H}_w$ correspond to non-leading terms in the large N_c expansion and are not factorizable, i.e., $\tilde{H}_w$ is responsible for non-factorizable amplitudes. The above can be described schematically in Fig. 1, where we have not explicitly shown soft gluon exchange(s) between quarks belonging to the same meson and the same current for simplicity. In the diagrams (a) and (c), no soft gluon(s) are exchanged between quarks belonging to different mesons or different currents. Therefore, the amplitudes described by these diagrams which correspond to the leading term in the large N_c expansion are factorizable. In these diagrams, the weak vertices are given by products of color singlet currents as in $H_w^{\rm BSW}$. However, the diagrams (b) and (d) correspond to non-leading terms and are non-factorizable since soft gluon(s) are exchanged between quarks belonging to different mesons or different currents, so that the weak vertices in these diagrams are given by products of colored currents as in $\tilde{H}_w$. In the large N_c argument, the latter is neglected while, in the color transparency, it is not guaranteed that the latter can be neglected. Therefore, $H_w^{\rm BSW}$ and $\tilde{H}_w$ are responsible for factorizable and non-factorizable amplitudes, respectively.

III. $\bar{B} \to D\pi$ AND $D^*\pi$ DECAYS

As discussed in the previous section, we start to study nonleptonic weak decays decomposing their amplitude into a sum of *factorizable* and *non-factorizable* ones [9,10]. The former is estimated by using the naive factorization in the BSW scheme [3]. The latter will be calculated later by using a hard pion approximation in the infinite momentum frame (IMF) [11,12]. The hard pion amplitude will be given by *asymptotic* matrix elements of $\tilde{H}_w$ (matrix elements of $\tilde{H}_w$ taken between single hadron states with infinite momentum).

Now we consider, as an example, the factorizable amplitude for the $B^-(p) \to D^0(p')\pi^-(q)$ decay. It is given by

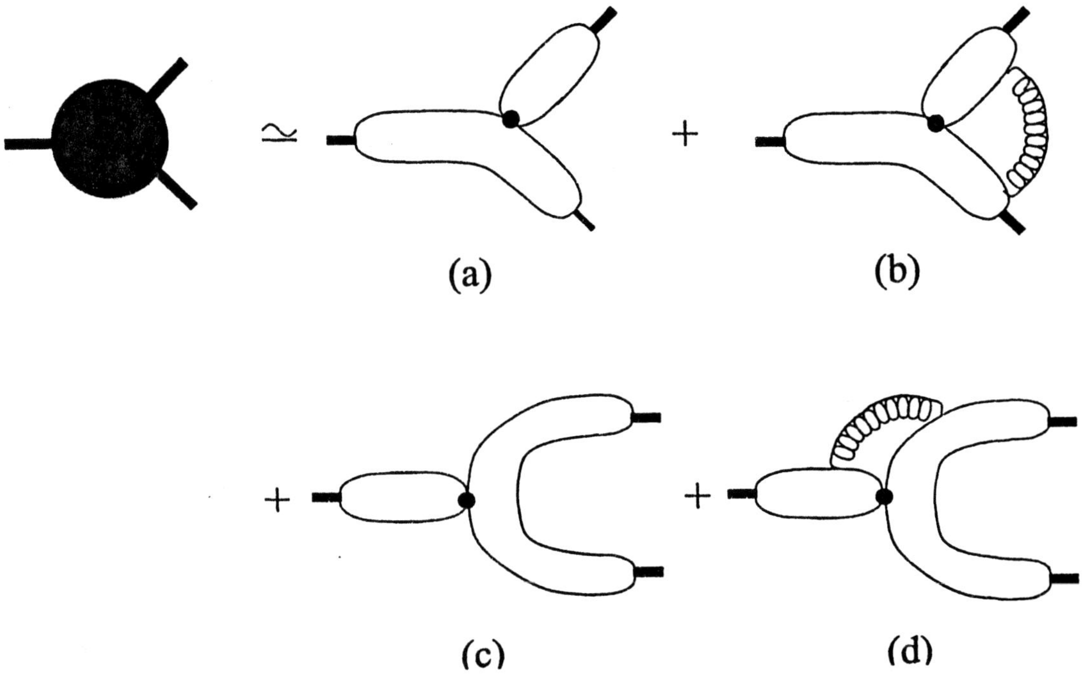

Fig. 1. Quark-line diagrams describing a hadronic weak decay of meson into two meson final states. Connectedness of quark-lines are assumed. The bullet denotes the weak vertex with perturbative QCD corrections and the spiral line denotes the soft-gluon(s). Soft-gluon exchange(s) between quarks belonging to the same meson and the same current are not explicitly described. The diagrams (a) and (c) correspond to the leading terms in the large N_c expansion while (b) and (d) to the non-leading terms.

$$\begin{aligned} &M_{\rm FA}(B^-(p) \to D^0(p')\pi^-(q)) \\ &\quad = \frac{G_F}{\sqrt{2}} V_{cb}V_{ud}\Big\{a_1\langle\pi^-(q)|(\bar{d}u)_{V-A}|0\rangle\langle D^0(p')|(\bar{c}b)_{V-A}|B^-(p)\rangle \\ &\qquad\qquad +a_2\langle D^0(p')|(\bar{c}u)_{V-A}|0\rangle\langle\pi^-(q)|(\bar{d}b)_{V-A}|B^-(p)\rangle\Big\}, \quad (7) \end{aligned}$$

in the BSW scheme. Factorizable amplitudes for the other $\bar{B} \to D\pi$ and $D^*\pi$ decays also can be calculated in the same way. To evaluate these amplitudes, we use the following parameterization of matrix elements of currents [4],

$$\langle\pi(q)|A_\mu|0\rangle = -if_\pi q_\mu, \qquad (8)$$

$$\langle D(p')|V_\mu|\bar{B}(p)\rangle = \left\{(p+p')_\mu - \frac{m_B^2 - m_D^2}{q^2}q_\mu\right\}F_1^{(D\bar{B})}(q^2)$$

$$+\frac{m_B^2 - m_D^2}{q^2} q_\mu F_0^{(D\bar{B})}(q^2), \tag{9}$$

$$\langle D^*(p')|A_\mu|\bar{B}(p)\rangle$$
$$= \left\{ (m_B + m_{D^*})\epsilon^*_\mu(p') A_1^{(D^*\bar{B})}(q^2) - \frac{\epsilon^*(p')\cdot q}{m_B + m_{D^*}}(p+p')_\mu A_2^{(D^*\bar{B})}(q^2) \right.$$
$$\left. -2m_{D^*}\frac{\epsilon^*\cdot q}{q^2} q_\mu A_3^{(D^*\bar{B})}(q^2) \right\} + 2m_{D^*}\frac{\epsilon^*\cdot q}{q^2} q_\mu A_0^{(D^*\bar{B})}(q^2), \tag{10}$$

where $q = p - p'$ and the form factors satisfy

$$A_3^{(D^*\bar{B})}(q^2) = \frac{m_B + m_{D^*}}{2m_{D^*}} A_1^{(D^*\bar{B})}(q^2) - \frac{m_B - m_{D^*}}{2m_{D^*}} A_2^{(D^*\bar{B})}(q^2), \tag{11}$$

$$F_1^{(D\bar{B})}(0) = F_0^{(D\bar{B})}(0), \qquad A_3^{(D^*\bar{B})}(0) = A_0^{(D^*\bar{B})}(0). \tag{12}$$

To get rid of useless imaginary unit in the amplitude, however, we take the following parameterization of matrix element of vector current [13],

$$\langle V(p')|V_\mu|0\rangle = -if_V m_V \epsilon^*_\mu(p'), \tag{13}$$

which can be treated in parallel to those of axial vector currents in Eq.(8) in the IMF. Using these expressions of current matrix elements, we obtain the factorized amplitudes for the $\bar{B} \to D\pi$ and $D^*\pi$ decays in Table 1, where we have put $m_\pi^2 = 0$ and factored out the CKM matrix elements. As seen in Table 1, the color suppressed amplitudes are proportional to the small coefficient, a_2, and therefore they are suppressed.

Before we evaluate numerically the factorized amplitudes, we study non-factorizale amplitudes for the $\bar{B}(p) \to D(p')\pi(q)$ [and $D^*(p')\pi(q)$] decays. The amplitudes for hadronic weak (two-body) decays with final state interactions will be described by diagrams similar to (b) and (c) in Fig. 1 in which soft gluon(s) are replaced by $\{q\bar{q}\}$ pair(s) [i.e., meson(s)], so that such amplitudes will be given by hadron dynamics and correspond to non-leading terms in the large N_c expansion as mentioned before. Therefore, we assume that the non-factorizable amplitudes are dominated by dynamical contributions of various hadron states, and then calculate them using a hard pion technique with PCAC (partially conserved axial-vector current) in the IMF; i.e., $\mathbf{p} \to \infty$ [11,12],

$$M_{\rm NF}(\bar{B} \to D^{[*]}\pi) \simeq \lim_{\mathbf{p}\to\infty,\ \mathbf{q}\to 0} M_{\rm NF}(\bar{B} \to D^{[*]}\pi). \tag{14}$$

In this approximation, $M_{\rm NF}$ is given by a sum of equal-time commutator (ETC) term and surface term,

Table 1. Factorized amplitudes for $\bar{B} \to D\pi$ and $D^*\pi$ decays where $m_\pi^2 = 0$. The CKM matrix elements are factored out.

Decay	$A_{\rm FA}$
$\bar{B}^0 \to D^+\pi^-$	$i\frac{G_F}{\sqrt{2}}a_1 f_\pi (m_B^2 - m_D^2) F_0^{(D\bar{B})}(0)\left[1 - \left(\frac{a_2}{a_1}\right)\left(\frac{f_B}{f_\pi}\right)\left(\frac{m_D^2}{m_B^2 - m_D^2}\right)\frac{F_0^{(D\pi)}(m_B^2)}{F_0^{(D\bar{B})}(0)}\right]$
$\bar{B}^0 \to D^0\pi^0$	$-i\frac{G_F}{2}a_2 f_D m_B^2 F_0^{(\pi\bar{B})}(m_D^2)\left[1 + \left(\frac{f_B}{f_D}\right)\left(\frac{m_D^2}{m_B^2}\right)\frac{F_0^{(D\pi)}(m_B^2)}{F_0^{(\pi\bar{B})}(m_D^2)}\right]$
$B^- \to D^0\pi^-$	$i\frac{G_F}{\sqrt{2}}a_1 f_\pi (m_B^2 - m_D^2) F_0^{(D\bar{B})}(0)\left[1 + \left(\frac{a_2}{a_1}\right)\left(\frac{f_D}{f_\pi}\right)\left(\frac{m_B^2}{m_B^2 - m_D^2}\right)\frac{F_0^{(\pi\bar{B})}(m_D^2)}{F_0^{(D\bar{B})}(0)}\right]$
$\bar{B}^0 \to D^{*+}\pi^-$	$-i\frac{G_F}{\sqrt{2}}a_1 f_\pi A_0^{(D^*\bar{B})}(0)\left[1 - \left(\frac{a_2}{a_1}\right)\left(\frac{f_B}{f_\pi}\right)\frac{A_0^{(D^*\pi)}(m_B^2)}{A_0^{(D^*\bar{B})}(0)}\right]2m_{D^*}\epsilon^*(p')\cdot p$
$\bar{B}^0 \to D^{*0}\pi^0$	$i\frac{G_F}{2}a_2 f_{D^*} F_1^{(\pi\bar{B})}(m_{D^*}^2)\left[1 + \left(\frac{f_B}{f_{D^*}}\right)\frac{A_0^{(D^*\pi)}(m_B^2)}{F_1^{(\pi\bar{B})}(m_{D^*}^2)}\right]2m_{D^*}\epsilon^*(p')\cdot p$
$B^- \to D^{*0}\pi^-$	$-i\frac{G_F}{\sqrt{2}}a_1 f_\pi A_0^{(D^*\bar{B})}(0)\left[1 + \left(\frac{a_2}{a_1}\right)\left(\frac{f_{D^*}}{f_\pi}\right)\frac{F_1^{(\pi\bar{B})}(m_{D^*}^2)}{A_0^{(D^*\bar{B})}(0)}\right]2m_{D^*}\epsilon^*(p')\cdot p$

$$M_{\rm NF}(\bar{B} \to D^{[*]}\pi) \simeq M_{\rm ETC}(\bar{B} \to D^{[*]}\pi) + M_{\rm S}(\bar{B} \to D^{[*]}\pi). \tag{15}$$

$M_{\rm ETC}$ has the same form as that in the old soft pion approximation [14]

$$M_{\rm ETC}(\bar{B} \to D^{[*]}\pi) = \frac{i}{f_\pi}\langle D^{[*]}|[V_{\bar{\pi}}, \tilde{H}_w]|\bar{B}\rangle. \tag{16}$$

In the above equation, the commutation relation, $[V_\pi + A_\pi, \tilde{H}_w] = 0$, has been used, where V_π and A_π are isospin charge and its axial counterpart, respectively. The surface term, M_S, is given by

$$M_{\rm S}(\bar{B} \to D^{[*]}\pi) = \lim_{\mathbf{p}\to\infty,\ \mathbf{q}\to 0}\Big\{-\frac{i}{f_\pi}q^\mu T_\mu^{[*]}\Big\}, \tag{17}$$

with the hypothetical amplitude

$$T_\mu^{[*]} = i\int e^{iqx}\langle D^{[*]}(p')|T[A_\mu^{(\bar{\pi})}\tilde{H}_w]|\bar{B}(p)\rangle d^4x. \tag{18}$$

When a complete set of energy eigen states is inserted between products of axial vector current and $\tilde{H}_w$ in $T_\mu^{[*]}$, contributions of single meson intermediate states survive in contrast with the soft pion approximation. In this way, $M_{\rm S}$ is provided by a sum of all possible pole amplitudes,

$$M_{\rm S} = \sum_n M_{\rm S}^{(n)} + \sum_l M_{\rm S}^{(l)}. \tag{19}$$

$M_{\rm S}^{(n)}$ and $M_{\rm S}^{(l)}$ are pole amplitudes in the s- and u-channels, respectively, and given by

$$\begin{aligned} M_{\rm S}^{(n)}(\bar{B} \to D^{[*]}\pi) &= -\frac{i}{f_\pi}\Big(\frac{m_{D^{[*]}}^2 - m_B^2}{m_n^2 - m_B^2}\Big)\langle D^{[*]}|A_{\bar{\pi}}|n\rangle\langle n|\tilde{H}_w|\bar{B}\rangle, \\ M_{\rm S}^{(l)}(\bar{B} \to D^{[*]}\pi) &= -\frac{i}{f_\pi}\Big(\frac{m_{D^{[*]}}^2 - m_B^2}{m_\ell^2 - m_{D^{[*]}}^2}\Big)\langle D^{[*]}|\tilde{H}_w|l\rangle\langle l|A_{\bar{\pi}}|\bar{B}\rangle. \end{aligned} \tag{20}$$

Therefore, $M_{\rm S}$ provides corrections to the soft pion approximation, i.e., the present hard pion technique can be considered as an innovation of the old soft pion technique. (See, for more details and notations, Refs. [11] and [12].) n and l in Eq.(20) run over all possible single meson states, not only ordinary $\{q\bar{q}\}$, but also hybrid $\{q\bar{q}g\}$, four-quark $\{qq\bar{q}\bar{q}\}$, glue-balls, etc. However, $|n\rangle$ and $|l\rangle$ as well as the external states are energy eigen states in the present case. Since the states which sandwich $\tilde{H}_w$ should conserve their spins in the rest frame and since Lorentz invariant amplitudes are considered, only the states $|n\rangle$ and $|l\rangle$ which conserve spins in the matrix elements, $\langle n|\tilde{H}_w|\bar{B}\rangle$ and $\langle D^{[*]}|\tilde{H}_w|l\rangle$, should be picked up [15]. Therefore, we discard vector meson pole contributions to the u-channel of pseudo-scalar(PS)-meson decays into two PS-meson states although we have taken such contributions in our previous papers [16]. Since the B meson mass m_B is much higher than those of charm mesons and since wave function overlappings between the ground-state $\{q\bar{q}\}_0$

and excited-state-meson states are expected to be small, however, excited meson contributions will be small in these decays and can be safely neglected. Therefore the hard pion amplitudes as the non-factorizable long distance ones are approximately described in terms of *asymptotic ground-state-meson matrix elements* (matrix elements taken between single ground-state-meson states with infinite momentum) of V_π, A_π and $\tilde{H}_w$.

Amplitudes for dynamical hadronic processes are usually decomposed into (*continuum contribution*) + (*Born term*). Since $M_{\rm S}$ has been given by a sum of pole amplitudes, $M_{\rm ETC}$ corresponds to the continuum contribution [17] which can develop a phase relative to the Born term. Therefore, we parameterize the ETC terms using isospin eigen amplitudes and their phases. Since the $D\pi$ final state can have isospin $I = \frac{1}{2}$ and $\frac{3}{2}$, $M_{\rm ETC}(\bar{B} \to D\pi)$'s are parameterized as

$$M_{\rm ETC}(\bar{B}^0 \to D^+\pi^-) = \sqrt{\frac{1}{3}}M^{(3)}_{\rm ETC}e^{i\tilde{\delta}_3} + \sqrt{\frac{2}{3}}M^{(1)}_{\rm ETC}e^{i\tilde{\delta}_1}, \tag{21}$$

$$M_{\rm ETC}(\bar{B}^0 \to D^0\pi^0) = -\sqrt{\frac{2}{3}}M^{(3)}_{\rm ETC}e^{i\tilde{\delta}_3} + \sqrt{\frac{1}{3}}M^{(1)}_{\rm ETC}e^{i\tilde{\delta}_1}, \tag{22}$$

$$M_{\rm ETC}(B^- \to D^0\pi^-) = \sqrt{3}M^{(3)}_{\rm ETC}e^{i\tilde{\delta}_3}, \tag{23}$$

where $M^{(2I)}_{\rm ETC}$'s are the isospin eigen amplitudes with isospin I and $\tilde{\delta}_{2I}$'s are the corresponding phase shifts introduced. In the present approach, therefore, the final state interactions are included in the non-factorizable amplitudes. This is compatible with the fact that amplitudes with final state interactions are given by diagrams which belong to non-leading terms in the large N_c expansion.

Asymptotic matrix elements of V_π and A_π are parameterized as

$$\begin{aligned}\langle\pi^0|V_{\pi^+}|\pi^-\rangle &= \sqrt{2}\langle K^+|V_{\pi^+}|K^0\rangle \\ &= -\sqrt{2}\langle D^+|V_{\pi^+}|D^0\rangle = \sqrt{2}\langle B^+|V_{\pi^+}|B^0\rangle = \cdots = \sqrt{2},\end{aligned} \tag{24}$$

$$\begin{aligned}\langle\rho^0|A_{\pi^+}|\pi^-\rangle &= \sqrt{2}\langle K^{*+}|A_{\pi^+}|K^0\rangle \\ &= -\sqrt{2}\langle D^{*+}|A_{\pi^+}|D^0\rangle = \sqrt{2}\langle B^{*+}|A_{\pi^+}|B^0\rangle = \cdots = h.\end{aligned} \tag{25}$$

The above parameterization can be obtained by using asymptotic $SU_f(5)$ symmetry [18], or $SU_f(5)$ extension of the nonet symmetry in $SU_f(3)$. Asymptotic matrix elements of V_π between vector meson states can be obtained by exchanging PS-mesons for vector mesons with corresponding flavors in Eq.(24), for example, as $\pi^{0,-} \to \rho^{0,-}$, etc. The $SU_f(4)$ part of the above parameterization reproduces well [12,19] the observed values of decay rates, $\Gamma(D^* \to D\pi)$ and $\Gamma(D^* \to D\gamma)$.

In this way, we can describe the non-factorizable amplitudes for the $\bar{B} \to D\pi$ decays as

$$M_{\rm NF}(\bar{B}^0 \to D^+\pi^-) \simeq -i\frac{\langle D^0|\tilde{H}_w|\bar{B}^0\rangle}{f_\pi}\left\{\left[\frac{4}{3}e^{i\tilde{\delta}_1} - \frac{1}{3}e^{i\tilde{\delta}_3}\right] + \cdots\right\}, \tag{26}$$

$$M_{\rm NF}(\bar{B}^0 \to D^0\pi^0) \simeq -i\frac{\langle D^0|\tilde{H}_w|\bar{B}^0\rangle}{f_\pi}\left\{\frac{\sqrt{2}}{3}\left[2e^{i\tilde{\delta}_1} + e^{i\tilde{\delta}_3}\right] + \cdots\right\}, \tag{27}$$

$$M_{\rm NF}(B^- \to D^0\pi^-) \simeq i\frac{\langle D^0|\tilde{H}_w|\bar{B}^0\rangle}{f_\pi}\left\{e^{i\tilde{\delta}_3} + \cdots\right\}, \tag{28}$$

where the ellipses denote the neglected pole contributions.

In the case of the $\bar{B} \to D^*\pi$ decays, the matrix element $\langle V|\tilde{H}_w|P\rangle$ should vanish because of conservation of spin as discussed before, so that $M_{\rm ETC}(\bar{B} \to D^*\pi)$ also should vanish but now D and B^* poles in the s- and u-channels, respectively, survive, i.e.,

$$M_{\rm NF}(\bar{B}^0 \to D^{*+}\pi^-) \simeq \frac{i}{f_\pi}\langle D^0|\tilde{H}_w|\bar{B}^0\rangle\left(\frac{m_B^2 - m_{D^*}^2}{m_B^2 - m_D^2}\right)\sqrt{\frac{1}{2}}h + \cdots, \tag{29}$$

$$M_{\rm NF}(\bar{B}^0 \to D^{*0}\pi^0) \simeq \frac{i}{\sqrt{2}f_\pi}\left[\langle D^0|\tilde{H}_w|\bar{B}^0\rangle\left(\frac{m_B^2 - m_{D^*}^2}{m_B^2 - m_D^2}\right)\right.$$
$$\left. + \langle D^{*0}|\tilde{H}_w|\bar{B}^{*0}\rangle\left(\frac{m_B^2 - m_{D^*}^2}{m_{B^*}^2 - m_{D^*}^2}\right)\right]\sqrt{\frac{1}{2}}h + \cdots, \tag{30}$$

$$M_{\rm NF}(B^- \to D^{*0}\pi^-) \simeq -\frac{i}{f_\pi}\langle D^{*0}|\tilde{H}_w|\bar{B}^{*0}\rangle\left(\frac{m_B^2 - m_{D^*}^2}{m_{B^*}^2 - m_{D^*}^2}\right)\sqrt{\frac{1}{2}}h + \cdots, \tag{31}$$

where the ellipses denote the neglected excited meson contributions. Therefore the non-factorizable amplitudes in the hard pion approximation are controlled by the asymptotic ground-state-meson matrix elements of $\tilde{H}_w$ (and the possible phases).

Now we evaluate the above amplitudes in the IMF. The factorized amplitudes in Table 1 contain many parameters which have not been measured by experiments, *i.e.*, form factors, $F_0^{(D\bar{B})}(q^2)$, $A_0^{(D^*\bar{B})}(q^2)$, $F_0^{(\pi\bar{B})}(q^2)$, $F_1^{(\pi\bar{B})}(q^2)$, etc., and decay constants, f_D, f_D^*, f_B, etc. The form factors $F_0^{(D\bar{B})}(0)$ and $A_0^{(D^*\bar{B})}(0)$ have been calculated by using the heavy quark effective theory (HQET) [20]. The other form factors are concerned with light mesons and therefore have to be estimated by some other models. In color favored decays, however, main parts of the factorized amplitudes depend on the form factor, $F_0^{(D\bar{B})}(0)$ or $A_0^{(D^*\bar{B})}(0)$, and the other form factors are included in minor terms proportional to a_2. Therefore our result may not be lead to serious uncertainties although some model dependent values of the form factors are taken. (We will take, later, $F_0^{(D\bar{B})}(0) \simeq A_0^{(D^*\bar{B})}(0) \simeq 0.59$ as in Ref. [21], $F_0^{(\pi\bar{B})}(m_D^2) \simeq 0.30$ and $F_1^{(\pi\bar{B})}(m_{D^*}^2) \simeq 0.34$ as expected in pQCD [22].) In the color suppressed $\bar{B}^0 \to D^0\pi^0$ and $D^{*0}\pi^0$ decays, however, the factorized amplitudes are proportional to the form factors, $F_0^{(\pi\bar{B})}(m_D^2)$ and $F_1^{(\pi\bar{B})}(m_{D^*}^2)$, respectively. Since their values are model dependent, the result on the color suppressed decays may be a little ambiguous, if non-factorizable contribution is less important. For the decay constants of heavy mesons, we assume $f_D \simeq f_{D^*}$ (and $f_B \simeq f_{B^*}$) since D and D^* (B and B^*) are expected to be degenerate because of heavy quark symmetry [20] and are approximately degenerate in reality. Here we take $f_{D^*} \simeq f_D \simeq 211$ MeV and $f_{B^*} \simeq f_B \simeq 179$ MeV from a recent result of lattice QCD [23]. In this way, we can obtain the factorized amplitudes in the second column of Table 2, where we have neglected very small annihilation terms in the $\bar{B}^0 \to D^0\pi^0$ and $D^{*0}\pi^0$ decay amplitudes.

To evaluate the non-factorizable amplitudes, we need to know the size of the asymptotic matrix elements of $\tilde{H}_w$ and A_π. The asymptotic matrix elements of A_π which was parameterized in Eq.(25) is estimated [11,12] to be $|h| \simeq 1.0$ by using PCAC and the observed rate [24], $\Gamma(\rho \to \pi\pi)_{\rm exp} \simeq 150$ MeV. We here take its positive sign, i.e., $h \simeq 1.0$, since, in the IMF, $\langle\rho^0|A_{\pi^+}|\pi^-\rangle$ is given by the form factor $A_3^{(\rho\pi)}(0)$ included in the matrix element of axial-vector current. In this way, the factorized and non-factorizable amplitudes for the $B^- \to D^{*0}\pi^-$ decay interfere constructively with each other. For the asymptotic matrix elements, $\langle D^0|\tilde{H}_w|\bar{B}^0\rangle$ and $\langle D^{*0}|\tilde{H}_w|\bar{B}^{*0}\rangle$, we treat them as unknown parameters and search phenomenologically for their values to reproduce the observed rates for the $\bar{B} \to D^{[*]}\pi$ decays. To this, we assume

$$\langle D^{*0}|\tilde{H}_w|\bar{B}^{*0}\rangle \simeq \langle D^0|\tilde{H}_w|\bar{B}^0\rangle \tag{32}$$

as expected in the heavy quark symmetry and parameterize these matrix elements using factorizable ones of $H_w^{\rm BSW}$ as

$$\langle D^0|\tilde{H}_w|\bar{B}^0\rangle = B_H\langle D^0|H_w^{\rm BSW}|\bar{B}^0\rangle_{\rm FA} \tag{33}$$

with

$$\langle D^0|H_w^{\rm BSW}|\bar{B}^0\rangle_{FA} = \frac{G_F}{\sqrt{2}}V_{cb}V_{ud}\Big(\frac{m_D^2 + m_B^2}{2}\Big)f_D f_B a_2, \tag{34}$$

where B_H is a parameter introduced. In this way, we obtain the hard pion amplitudes as the non-factorizable contributions listed in the third column of Table 2, where the CKM matrix elements have been factored out.

Table 2. Factorized and non-factorizable amplitudes for the $\bar{B} \to D\pi$ and $D^*\pi$ decays. The CKM matrix elements are factored out.

Decay	$A_{\rm FA}$ ($\times 10^{-5}$ GeV)	$A_{\rm NF}$ ($\times 10^{-5}$ GeV)
$\bar{B}^0 \to D^+\pi^-$	$1.57\, a_1\Big\{1 - 0.03\Big(\frac{a_2}{a_1}\Big)\Big\}$	$-3.70 a_2 B_H \Big\{\Big[\frac{4}{3} e^{i\tilde{\delta}_1} - \frac{1}{3} e^{i\tilde{\delta}_3}\Big]\Big\}$
$\bar{B}^0 \to D^0\pi^0$	$-1.03\, a_2\Big\{\frac{f_D}{0.211\ \text{GeV}}\Big\}$	$-3.70 a_2 B_H \Big\{\frac{\sqrt{2}}{3}\Big[2e^{i\tilde{\delta}_1} + e^{i\tilde{\delta}_3}\Big]\Big\}$
$B^- \to D^0\pi^-$	$1.57\, a_1\Big\{1 + 0.94\Big(\frac{a_2}{a_1}\Big)\Big\}$	$3.70 a_2 B_H \Big\{e^{i\tilde{\delta}_3}\Big\}$
$\bar{B}^0 \to D^{*+}\pi^-$	$-1.53\, a_1\Big\{1 - 0.22\Big(\frac{a_2}{a_1}\Big)\Big\}$	$-3.70 a_2 B_H \Big\{0.694\Big\}$
$\bar{B}^0 \to D^{*0}\pi^0$	$0.997\, a_2\Big\{\frac{f_{D^*}}{0.211\ \text{GeV}}\Big\}$	$3.70 a_2 B_H \Big\{0.983\Big\}$
$B^- \to D^{*0}\pi^-$	$-1.53\, a_1\Big\{1 + 0.92\Big(\frac{a_2}{a_1}\Big)\Big\}$	$-3.70 a_2 B_H \Big\{0.696\Big\}$

Before we compare our result with the observations, we study phenomenologically allowed branching ratios for the $\bar{B} \to D\pi$ and $D^*\pi$ decays. For the color suppressed $\bar{B}^0 \to D^0\pi^0$ and $D^{*0}\pi^0$ decays, in particular, only the upper limits of their branching ratios have been given as their measured values at the present stage. However, we can estimate their lower bounds compatible with the measured branching ratios for the charged modes. To this, we parameterize the amplitudes for these decays as

$$A(\bar{B}^0 \to D^{[*]+}\pi^-) = \sqrt{\frac{1}{3}} A_3^{[*]} e^{i\delta_3^{[*]}} + \sqrt{\frac{2}{3}} A_1^{[*]} e^{i\delta_1^{[*]}}, \qquad (35)$$

$$A(\bar{B}^0 \to D^{[*]0}\pi^0) = -\sqrt{\frac{2}{3}}A_3^{[*]}e^{i\delta_3^{[*]}} + \sqrt{\frac{1}{3}}A_1^{[*]}e^{i\delta_1^{[*]}}, \tag{36}$$

$$A(B^- \to D^{[*]0}\pi^-) = \sqrt{3}A_3^{[*]}e^{i\delta_3^{[*]}}, \tag{37}$$

since the $D\pi$ [and $D^*\pi$] final states can have $I = \frac{1}{2}$ and $\frac{3}{2}$, where $A_{2I}^{[*]}$'s and $\delta_{2I}^{[*]}$'s are isospin eigen amplitudes for the $\bar{B} \to D^{[*]}\pi$ decays and their phases, respectively. Taking positive values of the ratio of isospin eigen amplitudes, $r^{[*]} = A_3^{[*]}/A_1^{[*]}$, we obtain

$$\cos(\delta^{[*]}) = \Big(\frac{9R_{0-}^{[*]} - 1}{4}\Big)r^{[*]} - \frac{1}{r^{[*]}}, \tag{38}$$

where

$$\delta^{[*]} = \delta_1^{[*]} - \delta_3^{[*]} \quad \text{and} \quad r^{[*]} = \sqrt{\frac{1}{3R_{00}^{[*]}R_{0-}^{[*]} + 3R_{0-}^{[*]} - 1}}\,. \tag{39}$$

Table 3. Branching ratios (%) for $\bar{B} \to D\pi$ and $D^*\pi$ decays, where the values of the CKM matrix elements, $V_{cb} = 0.040$ and $V_{ud} = 0.97$, the lifetimes, $\tau(B^-) = 1.65 \times 10^{-12}$ s and $\tau(\bar{B}^0) = 1.55 \times 10^{-12}$ s from the updated experimental data [24], $a_1 = 1.024$ with the LO QCD corrections and phenomenological $a_2 = 0.21$, $B_H = 0.17$, $\tilde{\delta}_1 = 88°$ and $\tilde{\delta}_3 = 8°$ are taken. $\mathcal{B}_{\rm FA}$ and $\mathcal{B}_{\rm tot}$ are given by the factorized amplitude and a sum of the factorized and non-factorizable ones, respectively. The values of $\mathcal{B}_{\rm ph}$ are estimated in the text.

Decays	$\mathcal{B}_{\rm FA}$	$\mathcal{B}_{\rm tot}$	$\mathcal{B}_{\rm ph}$	$\mathcal{B}_{\rm exp}$ (*)
$\bar{B}^0 \to D^+\pi^-$	0.29	0.30	$0.28 - 0.34$	0.30 ± 0.04
$\bar{B}^0 \to D^0\pi^0$	0.005	0.011	$0.006 - 0.012$	< 0.012
$B^- \to D^0\pi^-$	0.45	0.52	0.53 ± 0.05	0.53 ± 0.05
$\bar{B}^0 \to D^{*+}\pi^-$	0.25	0.28	0.276 ± 0.021	0.276 ± 0.021
$\bar{B}^0 \to D^{*0}\pi^0$	0.005	0.013	$0.004 - 0.044$	< 0.044
$B^- \to D^{*0}\pi^-$	0.42	0.46	0.46 ± 0.04	0.46 ± 0.04

(*) The data values are taken from Ref. [24].

Here $R_{0-}^{[*]}$ and $R_{00}^{[*]}$ are ratios of decay rates,

$$R_{0-}^{[*]} = \frac{\Gamma(\bar{B}^0 \to D^{[*]+}\pi^-)}{\Gamma(B^- \to D^{[*]0}\pi^-)} \quad \text{and} \quad R_{00}^{[*]} = \frac{\Gamma(\bar{B}^0 \to D^{[*]0}\pi^0)}{\Gamma(\bar{B}^0 \to D^{[*]+}\pi^-)}. \tag{40}$$

Values of $R_{0-}^{[*]}$ and $R_{00}^{[*]}$ can be estimated phenomenologically from the experimental data [24] on branching ratios for $\bar{B} \to D^{[*]}\pi$ decays in Table 3 as $R_{0-} = 0.58 \pm 0.10$, $R_{00} < 0.04$ and $R_{0-}^* = 0.61 \pm 0.08$, $R_{00}^* < 0.16$. However, these values of R_{0-} and R_{00} [R_{0-}^* and R_{00}^*] are not always compatible with each other, i.e., the right-hand-side (r.h.s.) in Eq.(38) is not always less than unity. It is satisfied in more restricted regions of R and R^*, i.e., approximately, $0.04 \gtrsim R_{00} \gtrsim 0.02$ and $0.68 \gtrsim R_{0-} \gtrsim 0.61$ for the $\bar{B} \to D\pi$ decays, and $0.16 \gtrsim R_{00}^* \gtrsim 0.02$ and $0.69 \gtrsim R_{0-}^* \gtrsim 0.53$ for the $\bar{B} \to D^*\pi$ decays. These values of R and R^* lead approximately to the phenomenological branching ratios, $\mathcal{B}_{\rm ph}$'s in Table 3, when the experimental data, $\mathcal{B}(B^- \to D^0\pi^-)_{\rm exp} = 0.53 \pm 0.05$ and $\mathcal{B}(B^- \to D^{*0}\pi^-)_{\rm exp} = 0.46 \pm 0.04$, are fixed. Here we put $\mathcal{B}(\bar{B}^0 \to D^{*+}\pi^-)_{\rm ph} = \mathcal{B}(\bar{B}^0 \to D^{*+}\pi^-)_{\rm exp}$ since $\cos(\delta^*) \leq 1$ is satisfied for all the experimentally allowed values of R_{0-}^*. The allowed values of $\cos(\delta)$ are limited within a narrow region, $0.96 \lesssim \cos(\delta) \leq 1$, in which $\cos(\delta)$ is very close to unity while $\cos(\delta^*)$ is a little more mildly restricted (at the present stage) compared with the above $\cos(\delta)$, i.e., approximately, $0.70 \lesssim \cos(\delta^*) \leq 1$. Therefore, the phase difference between the $I = \frac{1}{2}$ and $\frac{3}{2}$ amplitudes for the $\bar{B} \to D\pi$ decays is very small while, in the $\bar{B} \to D^*\pi$ decays, the allowed region of the corresponding phase difference is broader at the present stage.

We now compare our result on the branching ratios, $\mathcal{B}(\bar{B} \to D\pi)$ and $\mathcal{B}(\bar{B} \to D^*\pi)$, with their phenomenologically estimated values (and experimental data), taking a sum of the factorized amplitude and the non-factorizable amplitude as the total one. To this, we determine values of parameters involved. We take $V_{cb} = 0.040$ from the updated value $|V_{cb}| = 0.0402 \pm 0.0019$ [24]. For the coefficients a_1 and a_2 in $H_w^{\rm BSW}$, we do not know their true values. It has been known [25] that the NLO corrections to a_1 are small while the corresponding ones to a_2 may be not much smaller compared with the LO corrections and depend strongly on the renormalization scheme. Therefore, we expect that the value, $a_1 = 1.024$ with the LO corrections at the scale $\mu \simeq m_b$ [25], is not very far from the true value, and hence we take the above value of a_1. For a_2, however, we consider two cases. As the first case (i), we take $a_2 = 0.125$ with the LO QCD corrections at $\mu \simeq m_b$ [25], and then we treat it as an adjustable parameter around the above value of a_2 as the second case (ii). For the phases $\tilde{\delta}_1$ and $\tilde{\delta}_3$ arising from contributions of non-resonant multi-hadron intermediate states into isospin $I = \frac{1}{2}$ and $\frac{3}{2}$ final states, they are restricted in the region $|\tilde{\delta}_{2I}| < 90°$ since resonant contributions have already been extracted as pole amplitudes in M_S although their contributions are neglected as discussed before. For B_H, we here treat it as a free parameter but expect to be less than unity.

We now search for values of parameters, $\tilde{\delta}_1$, $\tilde{\delta}_3$ and B_H in the case (i), and a_2, $\tilde{\delta}_1$, $\tilde{\delta}_3$ and B_H in the case (ii), which reproduce the phenomenologically estimated branching ratios (from the observed ones) for the $\bar{B} \to D^{[*]}\pi$ decays. In the case (i), it is hard to reproduce them simultaneously. In the case (ii), however, large $\tilde{\delta}_1$, $(90° \succ \tilde{\delta}_1 \gtrsim 60°)$, and small $|\tilde{\delta}_3|$ are favored (but our result is not very sensitive to the latter). For a_2 and B_H, larger values of a_2, $(0.26 \gtrsim a_2 \gtrsim 0.16)$, compared with $a_2 = 0.125$ with the LO QCD corrections and rather small values of B_H, $(0.07 \lesssim B_H \lesssim 0.28)$, are favored. We list our results on the branching ratios for $a_1 = 1.024$, $a_2 = 0.21$, $\tilde{\delta}_1 = 88°$, $\tilde{\delta}_3 = 8°$ and $B_H = 0.17$ in Table 3, where we have used the values, $V_{cb} = 0.040$, $V_{ud} = 0.97$, $\tau(B^-) = 1.65 \times 10^{-12}$ s and $\tau(\bar{B}^0) = 1.55 \times 10^{-12}$ s from the updated experimental data [24]. $\mathcal{B}_{\rm FA}$ and $\mathcal{B}_{\rm tot}$ are given by the factorized amplitude and a sum of the factorized and non-factorizable ones, respectively. Values of $\mathcal{B}_{\rm ph}$ have been obtained phenomenologically from $\mathcal{B}_{\rm exp}$ before. $\mathcal{B}_{\rm FA}$, in which the non-factorizable contributions are discarded, reproduces fairly well the existing data. However, if we add the non-factorizable contributions, we can improve the fit to the phenomenologically estimated $\mathcal{B}_{\rm ph}$. It is seen that the non-factorizable contributions to the color favored $\bar{B} \to D\pi$ and $D^*\pi$ decays are rather small but still can interfere efficiently with the main amplitude given by the naive factorization.

IV. $\bar{B} \to J/\psi\bar{K}$ AND $J/\psi\pi$ DECAYS

Now we study CKM-angle favored $\bar{B} \to J/\psi\bar{K}$ and suppressed $\bar{B} \to J/\psi\pi$ decays in the same way as in the previous section. Both of them are color mismatched and their kinematical condition is much different from the color favored $\bar{B} \to D\pi$ and $D^*\pi$ decays at the level of underlying quarks, i.e., $b \to (c\bar{c})_1 + s$ (or d) in the former but $b \to c + (\bar{u}d)_1$ in the latter. Therefore, if the large N_c argument does not work well in hadronic weak decays as seen before, dominance of factorized amplitudes in these decays cannot be guaranteed and hence non-factorizable long distance contribution can play an important role.

The factorized amplitude for the $\bar{B} \to J/\psi\bar{K}$ decays is given by

$$M_{\rm FA}(\bar{B} \to J/\psi\bar{K}) = -iV_{cb}V_{cs}\Big\{\frac{G_F}{\sqrt{2}}a_2 f_\psi F_1^{(\bar{K}\bar{B})}(m_\psi^2)\Big\}2m_\psi\epsilon^*(p')\cdot p. \qquad (41)$$

The value of the decay constant of J/ψ is estimated to be $f_\psi \simeq 406$ MeV from the observed rate [24] for the $J/\psi \to \ell^+\ell^-$. The value of the CKM matrix element V_{cs} is given by $V_{cs} \simeq V_{ud} \simeq 0.97$. The form factor $F_1^{(\bar{K}\bar{B})}(m_\psi^2)$ has not been measured and its theoretical estimates are model dependent. We pick out tentatively three typical values of $F_1^{(\bar{K}\bar{B})}(m_\psi^2)$ by the models, BSW [3], CDDFGN [26] and GKP [27], among many models and list the resulting $\mathcal{B}_{\rm FA}(\bar{B} \to J/\psi\bar{K})$ in Table 4, where we have used the same values of parameters as before. $\mathcal{B}_{\rm FA}$ from the factorized amplitude for the value of $F_1^{(\bar{K}\bar{B})}(m_\psi^2)$ by BSW, which is close to the most recent value from pQCD [22] with $SU_f(3)$ symmetry, is about a half of the observations [24], $\mathcal{B}(B^- \to J/\psi K^-)_{\rm exp} = (0.100 \pm 0.010)$ % and $\mathcal{B}(\bar{B}^0 \to J/\psi\bar{K}^0)_{\rm exp} = (0.089 \pm 0.012)$ %. It means that some other (non-factorizable) contribution is needed in this case. On the other hand, $\mathcal{B}_{\rm FA}$'s for the higher values (CDDFGN and GKP) of the form factor saturate the above data values even if non-factorizable contributions are not included. It will be seen later that the predicted values of $\mathcal{B}_{\rm tot}$ including non-factorizable contributions is a little too large, in particular, in the case of GKP.

Table 4. Branching ratios (%) for the $\bar{B} \to J/\psi\bar{K}$ decays, where the values of $F_1^{(\bar{K}\bar{B})}(m_\psi^2)$ from the models, BSW, CDDFGN and GKP, in Refs. [3], [26] and [27], respectively, are used. Values of the other parameters involved are the same as in Table 3, where $B'_H = B_H$ is assumed. The data values are taken from Ref. [24].

Models	BSW	CDDFGN	GKP
$F_1^{(\bar{K}\bar{B})}(m_\psi^2)$	0.565	0.726	0.837
$\mathcal{B}_{\rm FA}(\bar{B} \to J/\psi\bar{K})$ (%)	0.048	0.079	0.10
$\mathcal{B}_{\rm tot}(\bar{B} \to J/\psi\bar{K})$ (%)	0.075	0.11	0.14
Experiments	$\mathcal{B}(B^- \to J/\psi K^-) = (0.100 \pm 0.010)$ % $\mathcal{B}(\bar{B}^0 \to J/\psi\bar{K}^0) = (0.089 \pm 0.012)$ %		

Non-factorizable contributions to these decays are estimated by using a hard kaon approximation which is a simple extension of the hard pion technique in the previous section. With this approximation and isospin symmetry, non-factorizable amplitude for the $\bar{B} \to J/\psi\bar{K}$ decays is given by

$$M_{\rm NF}(\bar{B} \to J/\psi\bar{K}) = \frac{i}{f_K}\langle\psi|H_w|\bar{B}_s^{*0}\rangle\left(\frac{m_B^2 - m_\psi^2}{m_{B_s^*}^2 - m_\psi^2}\right)\sqrt{\frac{1}{2}}h + \cdots, \qquad (42)$$

where the ellipsis denotes neglected contributions of excited mesons [28]. Here we have used $\langle\bar{B}_s^0|V_{K^+}|B^-\rangle = -1$ and $\sqrt{2}\langle\bar{B}_s^{*0}|A_{K^+}|B^-\rangle = -h$ which are flavor $SU_f(3)$ extensions of Eqs.(24) and (25). Asymptotic matrix element, $\langle\psi|\tilde{H}_w|\bar{B}_s^{*0}\rangle$, is parameterized in a way similar to that of $\langle D^{*0}|\tilde{H}_w|\bar{B}^{*0}\rangle$, i.e., $\langle\psi|\tilde{H}_w|\bar{B}_s^{*0}\rangle = B'_H|\langle\psi|H_w^{\rm BSW}|\bar{B}_s^{*0}\rangle_{\rm FA}|$, where B'_H is a parameter corresponding to B_H in Eq. (33). Then the total amplitude for the $\bar{B} \to J/\psi\bar{K}$ decays is approximately given by

$$M_{\rm tot}(\bar{B} \to J/\psi\bar{K}) \simeq -iV_{cb}V_{cs}\{6.12F_1^{(\bar{K}\bar{B})}(m_\psi^2) + 5.27B'_H\}a_2 \times 10^{-5}\ {\rm GeV}, \tag{43}$$

where $f_K \simeq 160$ MeV and $f_{B_s^*} \simeq f_{B_s} \simeq 204$ MeV from the updated lattice QCD result [23] have been taken.

When we take $a_2 = 0.21$ as before and assume tentatively $B'_H = B_H$ ($= 0.17$ as before), our result, $\mathcal{B}_{\rm tot}$, for the smaller value of the form factor, $F_1^{(\bar{K}\bar{B})}(m_\psi^2)$, from BSW [or pQCD with $SU_f(3)$ symmetry] is improved considerably. Contrary, $\mathcal{B}_{\rm tot}$ for the larger values of the form factor, in particular, from GKP is beyond the measured ones as seen in Table 4.

For the CKM-angle suppressed $\bar{B} \to J/\psi\pi$, the same technique and values of parameters as the above lead to

$$\begin{aligned} M_{\rm tot}(B^- \to J/\psi\pi^-) &\simeq -\sqrt{2}M_{\rm tot}(\bar{B}^0 \to J/\psi\pi^0) \\ &\simeq -iV_{cb}V_{cd}\{6.12F_1^{(\pi\bar{B})}(m_\psi^2) + 6.64B'_H\}a_2 \times 10^{-5}\ {\rm GeV}. \end{aligned} \tag{44}$$

The first equation is consistent with the measurements, $\mathcal{B}(B^- \to J/\psi\pi^-)_{\rm exp} = (5.1 \pm 1.5) \times 10^{-5}$ [24] and $\mathcal{B}(\bar{B}^0 \to J/\psi\pi^0) = (2.5^{+1.1}_{-0.9} \pm 0.2) \times 10^{-5}$ [29]. Using $F_1^{(\pi\bar{B})}(m_\psi^2) \simeq 0.56$ estimated by pQCD [22], which is close to the one of $F_1^{(\bar{K}\bar{B})}(m_\psi^2)$ from BSW, we obtain

$$\mathcal{B}_{\rm tot}(B^- \to J/\psi\pi^-) \simeq 4.2 \times 10^{-5}, \tag{45}$$

which should be compared with the the above measurement.

As seen in Table 4, our result on the color suppressed $\bar{B} \to J/\psi\bar{K}$ and $\bar{B} \to J/\psi\pi$ decays is still sensitive to the values of both of $F_1^{(\pi\bar{B})}(m_\psi^2)$ [or $F_1^{(\bar{K}\bar{B})}(m_\psi^2)$] and B'_H. It implies that both of the factorized and non-factorizable amplitudes are important in these decays.

V. SUMMARY

In the previous sections, we have studied the $\bar{B} \to D\pi$, $D^*\pi$, $J/\psi\bar{K}$ and $J/\psi\pi$ decays describing their amplitude by a sum of factorizable and non-factorizable ones. The former has been estimated by using the naive factorization while the latter has been calculated by using a hard pion (or kaon) approximation in the infinite momentum frame. The so-called final state interactions (corresponding to non-leading terms in the large N_c expansion) have been included in the non-factorizable long distance contributions. The corresponding ones in the color favored $\bar{B} \to D\pi$ and $D^*\pi$ decays are rather small and therefore the final state interactions seem to be not very important in these decays although still not necessarily negligible. Next, we have investigated phenomenologically the $\bar{B} \to D\pi$ and $D^*\pi$ decays, and observed that the existing data on their branching ratios are not always compatible with each other, i.e., $\cos(\delta^{[*]})$ is over unity for some values of $R_{00}^{[*]}$ and $R_{0-}^{[*]}$, where $\delta^{[*]}$ denote phase differences between amplitudes for the decays into $I = \frac{1}{2}$ and $\frac{3}{2}$ $D^{[*]}\pi$ final states. Eliminating such values of $R_{00}^{[*]}$ and $R_{0-}^{[*]}$, we have obtained phenomenologically allowed values of branching ratios, $\mathcal{B}_{\rm ph}$, in Table 3, which keep approximately $\cos(\delta^{[*]}) < 1$ and include lower limits on $\mathcal{B}(\bar{B} \to D^0\pi^0)$ and $\mathcal{B}(\bar{B} \to D^{*0}\pi^0)$.

By taking $a_1 \simeq 1.024$ with the LO QCD corrections and the phenomenological $a_2 \simeq 0.21$ which has been suggested previously [21,30], the observed branching ratios for these decays can be well reproduced in terms of a sum of the hard pion amplitude and the factorized one. Namely, the factorized amplitudes are dominant but not necessarily complete and long distance hadron dynamics should be carefully taken into account in hadronic weak interactions of B mesons. In color suppressed $\bar{B}^0 \to D^0\pi^0$, $D^{*0}\pi^0$, $\bar{B} \to J/\psi\bar{K}$ and $J/\psi\pi^-$ decays, non-factorizable long distance contributions are more important. In particular, in the $\bar{B} \to J/\psi\bar{K}$ decays, long distance physics should be treated carefully. When $a_2 \simeq 0.125$ with the LO QCD corrections is taken instead of the phenomenological $a_2 \simeq 0.21$, it is hard to reproduce the measured $\mathcal{B}(\bar{B} \to J/\psi\bar{K})_{\rm exp}$ and $\mathcal{B}(B^- \to J/\psi\pi^-)_{\rm exp}$ even if a sum of factorized and non-factorizable amplitudes is taken as long as $B'_H = B_H \simeq 0.17$.

The non-factorizable amplitudes are proportional to asymptotic ground-state-meson matrix elements of $\tilde{H}_w$, i.e., B_H or B'_H. To reproduce the measured rates for the color favored $\bar{B} \to D\pi$ and $D^*\pi$ decays, the non-factorizable contributions may not be negligible ($B_H \neq 0$) while too large values of B_H and B'_H will lead to too large rates for the color suppressed decays. However, their numerical results are still ambiguous since the amplitudes for the color suppressed decays are sensitive to model dependent form factors.

Experimental data on exclusive decays, in particular, decays into charmless final states, of B mesons are rapidly increasing after the B factories. In these decays, the factorization will not be a good approximation since the large N_c argument does not work in hadronic weak processes and since these decays are outside of applicability of the color transparency, so that non-factorizable contributions are expected to be very important in these decays. Besides, some of these decays are expected to play important roles in determination of CP-violating parameters. Therefore, non-factorizable contribution in exclusive decays of B mesons is now and will be, in near future, one of very important subjects.

The author would like to thank Prof. S. Pakvasa, Prof. H. Yamamoto, Prof. S. F. Tuan and the other members of high energy physics group, University of Hawaii for their discussions, comments and hospitality during his stay there, and Dr. D.-X. Zhang for comments. He also would like to appreciate Prof. T. Kurimoto for sending numerical values of form factors by pQCD.

References

[1] The factorization prescription was first proposed in S. Oneda and A. Wakasa, Nucl. Phys. **1**, 445 (1956) and then S. Oneda, J. C. Pati and B. Sakita, Phys. Rev. **119**, 482 (1960) in the theoretical frame work of the Sakata model in S. Sakata, Prog Theor. Phys. **16**, 686 (1956).
[2] M. A. Shifman, A. I. Vainstein, and V. I. Zakharov, Nucl. Phys. **B120**, 316 (1977).
[3] M. Bauer, B. Stech and M. Wirbel, Z. Phys. C **34**, 103 (1987).
[4] M. Neubert, V. Rieckert, B. Stech, and Q. P. Xu, in *Heavy Flavours*, edited by A. J. Buras and M. Lindner (World Scientific, Singapore, 1992).
[5] M. Fukugita, T. Inami, N. Sakai, and S. Yazaki, Phys. Lett. **72B**, 237 (1977); A. J. Buras, J. -M. Gerard, and R. Rückl, Nucl. Phys. **B268**, 16 (1986).
[6] J. D. Bjorken, Nucl. Phys. **B** (Proc. Suppl.) **11**, 325 (1989); M. J. Dugan and B. Grinstein, Phys. Lett. **255B**, 583 (1991).
[7] T. E. Browder, K. Honsheid and D. Pedrini, Ann. Rev. Nucl. Part. Sci. **46**, 395 (1996).
[8] N. Cabibbo, Phys. Rev. Lett. **10**, 531 (1963); M. Kobayashi and T. Maskawa, Prog. Theor. Phys. **49**, 652 (1973).
[9] K. Terasaki, Phys. Rev. D **54**, 3649 (1996); Int. J. Mod. Phys. A **13**, 4325 (1998).
[10] K. Terasaki, Phys. Rev. D **59**, 114001 (1999).
[11] K. Terasaki, S. Oneda, and T. Tanuma, Phys. Rev. D **29**, 456 (1984).
[12] S. Oneda and K. Terasaki, Prog. Theor. Phys. Suppl. **82**, 1 (1985).
[13] G. D. Haas and M. Youssefmir, Phys. Lett. **272B**, 391 (1991).
[14] H. Sugawara, Phys. Rev. Lett. **15**, 870 and 997 (E) (1965); M. Suzuki, *ibid.*, 986 (1965).

[15] S. Pakvasa, private communication (1999).
[16] Refs. [9–12] and references quoted therein.
[17] V. S. Mathur and L. K. Pandit, *Advances in Particle Physics*, edited by R. L. Cool and R. E. Marshak (Interscience Publishers, New York, 1968), Vol. 2, p. 383.
[18] Asymptotic flavor symmetry was proposed in S. Matsuda and S. Oneda, Phys. Rev. **158**, 1594 (1967), and was formulated in S. Oneda, H. Umezawa, and S. Matsuda, Phys. Rev. Lett. **25**, 71 (1970), and S. Oneda and S. Matsuda, Phys. Rev. D **2**, 324 (1970). A comprehensive review on the fruitful results from the asymptotic flavor symmetry has been provided in Ref. [12].
[19] E. Takasugi and S. Oneda, Phys. Rev. Lett. **34**, 1129 (1975); Phys. Rev. D **12**, 198 (1975); H. Hallock, S. Oneda, and M. D. Slaughter, *ibid.* **15**, 884 (1977); S. Oneda and Y. Koide, *Asymptotic Symmetry and Its Implication in Elementary Particle Physics* (World Scientific, Singapore, 1991).
[20] For a review on the heavy quark effective theory, see, for example, N. Isgur and M. B. Wise, in *Heavy Flavours*, edited by A. J. Buras and M. Lindner, Advance Series on Directions in High Energy Physics (World Scientific, Singapore, 1992), p. 234.
[21] A. N. Kamal and T. N. Pham, Phys. Rev. D **50**, 395 (1994); M. Gourdin, A. N. Kamal, Y. Y. Keum, and X. Y. Pham, Phys. Lett. **333B**, 507 (1994).
[22] T. Kurimoto, private communication (2000). For $F_1^{(\bar{K}\bar{B})}(m_\psi^2)$, we have assumed $F_1^{(\bar{K}\bar{B})}(m_\psi^2) \simeq F_1^{(\pi\bar{B})}(m_\psi^2)$ expected from the flavor $SU_f(3)$ symmetry.
[23] D. Becirevic *et al.*, hep-lat/9811003.
[24] Particle Data Group, D. E. Groom *et al.*, Eur. Phys. J. C **15**, 1 (2000).
[25] A. J. Buras, Nucl. Phys. **B434**, 606 (1995); G. Buchalla, A. J. Buras and M. E. Lautenbacher, Rev. Mod. Phys. **68**, 1125 (1996).
[26] R. Casalbouni, A. Deandrea, N. Di Bartolomeo, R. Gatto, F. Feruglio and G. Nardulli, Phys. Lett. **299B**, 139 (1993); A. Deandrea, N. Di Bartolomeo and R. Gatto, Phys. Lett. **318B**, 549 (1993).
[27] M. Gourdin, A. N. Kamal and X. Y. Pham, Phys. Rev. Lett. **73**, 3355 (1994).
[28] The $\bar{B} \to J/\psi\bar{K}$ decays are described by the color mismatched spectator diagram $b \to (c\bar{c})_1 + s$. Therefore the intermediate states in the s-channel of these decays are exotic, *i.e.*, $\{c\bar{c}s\bar{q}\}$, where $q = u$ or d is a spectator, and hence only four quark mesons can contribute as meson poles in the s-channel if the connectedness of quark lines is assumed as usual. However, masses of scalar $\{c\bar{c}s\bar{q}\}$, ($q = u$ or d), mesons are expected to be around 4

GeV and wave function overlapping between the four quark meson and the ground-state B meson states will be small. Therefore their contributions will be not very strongly enhanced. It is in contrast with the paper by F. E. Close, I. Dunietz, P. R. Page, S. Veseli and H. Yamamoto, Phys. Rev. D **57**, 5653 (1998) in which importance of contributions of hybrid mesons with masses around 4 GeV to decays described by the same type of quark-line diagrams has been stressed.
[29] CLEO Collaboration, P. Avery *et al.*, CLNS 00/1671, CLEO 00-8, hep-ex/0004032.
[30] M. S. Alam *et al.*, CLEO Collaboration, Phys. Rev. D **50**, 43 (1994).

Theoretical Physics 2002, Part 1
Edited by Thomas F. George and Henk F. Arnoldus, pp. 123-132
ISBN 1-59033-435-3

Central Configurations for n-Body Problems*

Zhang Shiqing†

Abstract

This paper surveys some joint works of the author with Xie and Zhou for central configurations of n-body problems.

1 Introduction

The Newtonian n-body problem [10] concerns the motion of n point particles with masses $m_j \in R^+$ and positions $\tilde{q}_j \in R^l (l = 2, 3; j = 1, \cdots, n)$. The motion is governed by Newton's second law and gravitational law:

$$m_j \ddot{\tilde{q}}_j = \frac{\partial U(\tilde{q})}{\partial \tilde{q}_j}, \tag{1.1}$$

where $\tilde{q} = (\tilde{q}_1, \cdots, \tilde{q}_n)$ and $U(\tilde{q})$ is the Newtonian potential:

$$U(\tilde{q}) = \sum_{1 \leq k < j \leq n} \frac{m_k m_j}{|\tilde{q}_k - \tilde{q}_j|}. \tag{1.2}$$

Consider the space

$$X = \{\tilde{q} = (\tilde{q}_1, \cdots, \tilde{q}_n) \in R^{ln} : \sum_{k=1}^{n} m_k \tilde{q}_k = 0\},$$

i.e., suppose that the center of mass is fixed at the origin of the space. Because the potential is singular when two particles have the same position, it is natural to assume that the configuration avoids the set $\triangle = \{\tilde{q} : \tilde{q}_k = \tilde{q}_j \text{ for some } k \neq j\}$. The set $X \backslash \triangle$ is called the configuration space.

Definition 1.1. [10] A configuration $\tilde{q} = (\tilde{q}_1, \cdots, \tilde{q}_n) \in X \backslash \triangle$ is called a central configuration (c.c.) if there exists a constant λ such that

$$\sum_{j=1, j\neq k}^{n} \frac{m_j m_k}{|\tilde{q}_j - \tilde{q}_k|^3} (\tilde{q}_j - \ tildeq_k) = -\lambda m_k \tilde{q}_k, \qquad 1 \leq k \leq n. \tag{1.3}$$

---

*Supported by NSF of China
†Department of Mathematics, College of Science, Chongqing University, Chongqing 400044, P.R.China

The value of the constant λ in (1.3) is uniquely determined by

$$\lambda = \frac{U}{I}, \tag{1.4}$$

where $I = \sum_{k=1}^{n} m_k |\tilde{q}_k|^2$.

2 Central Configurations for Planar N-Body Problems

This section uses the same notations as in [8]. For $N \geq 2$ the equations of motion of the planar N-body problem can be written in the form

$$\ddot{z}_k = -\sum_{j \neq k} m_j \frac{z_k - z_j}{|z_k - z_j|^3}, \tag{2.1}$$

where z_k is the complex coordinate of the $k-th$ mass m_k in an inertial coordinate system. In equation (2.1) and throughout this section, unless otherwise restricted, all indices and summations will range from 1 to N. Let ρ_k denote the N complex $k-th$ roots of unity, i.e.,

$$\rho_k = \exp(2\pi i k/N). \tag{2.2}$$

This equation will also serve to define ρ_k for any number k. The center of N masses m_k located at the vertices ρ_k of a regular polygon inscribed on the unit circle, is then given by

$$z_0 = \sum_j m_j \rho_j / M, \tag{2.3}$$

where $M = \sum_j m_j$. The functions describing their rotation about z_0 with angular velocity ω are then given by

$$z_k(t) = (\rho_k - z_0)\exp(i\omega t). \tag{2.4}$$

There is no loss of generality in assuming that the regular polygon is inscribed in the unit circle since the N-body problem (2.1) is invariant under the transformation $t \to t/a^{3/2}, z \to z/a$ which reduces the functions $a(\rho_k - z_0)\exp(i\omega t)$ to the functions defined in (2.4).

Xie-Zhang [11] gave simpler proofs of the following results of Perko-Walter-Elmabsout [8,3].

Theorem 2.1(Perko-Walter, [8]) If, for $N \geq 2$, the functions $z_k(t)$ given by (2.4) are solutions of the N-body problem (2.1), it follows that $\omega^2 = M\gamma/N$ where

$$\gamma = \frac{1}{4} \sum_{j \neq N} \csc(\pi j/N). \tag{2.5}$$

Corollary 2.1 (Lagrange, [5]) For N=3, the functions $z_k(t)$ given by (2.4) are solutions of the 3-body problem (2.1) if and only if $\omega^2 = M/3\sqrt{3}$.

Theorem 2.2(Perko-Walter, [8], Elmabsout, [3]) For $N \geq 4$ and $m_k > 0$ the functions $z_k(t)$ given by (2.4) with $\omega^2 = M\gamma/N$ and γ given by (2.5) are solutions of the N-body problem (2.1) if and only if $m_1 = m_2 = \cdots = m_N$.

3 Central Configurations for Planar 2N-Body Problems

We use the same notations as in paper [8]. For $N \geq 2$ the equations of motion of the planar 2N-body problem [8,10] can be written in the form

$$\ddot{z}_k = -\sum_{1 \leq j \neq k}^{2N} m_j \frac{z_k - z_j}{|z_k - z_j|^3}, \qquad k = 1, \cdots, 2N, \tag{3.1}$$

where z_k is the complex coordinate of the $k-th$ mass m_k in an inertial coordinate system. In the following all indices and summations will range from 1 to N. Let ρ_k denote the N complex $k-th$ roots of unity, i.e.,

$$\rho_k = \exp(2\pi i k/N). \tag{3.2}$$

This equation will also serve to define ρ_k for any number k. We assume masses $m_k (k = 1, \cdots, N)$ are located at the vertices ρ_k of a regular polygon inscribed on the unit circle, and $\tilde{m}_k = bm_k (b > 0, k = 1, \cdots, N)$ are located at another regular polygon:

$$\tilde{\rho}_k = a\rho_k e^{i\varphi}, \tag{3.3}$$

where $a > 0, 0 \leq \varphi \leq 2\pi$, and when $\varphi = 0$ or $2\pi, a \neq 1$. Then the center of the masses $m_1, \cdots, m_N, \tilde{m}_1, \cdots, \tilde{m}_N$ is

$$z_0 = \sum_j (m_j \rho_j + \tilde{m}_j \tilde{\rho}_j)/M, \tag{3.4}$$

where $M = \sum_j (m_j + \tilde{m}_j)$. The functions describing their rotation about z_0 with angular velocity ω are then given by

$$z_k(t) = (\rho_k - z_0)\exp(i\omega t), \qquad k = 1, \cdots, N, \tag{3.5}$$

and

$$\tilde{z}_k(t) = (a\rho_k e^{i\varphi} - z_0)\exp(i\omega t), \qquad k = 1, \cdots, N. \tag{3.6}$$

Then the equations of motion of the planar 2N-body problem can be written in the form

$$\ddot{z}_k = \sum_{j \neq k} m_j \frac{z_j - z_k}{|z_j - z_k|^3} + \sum_j \tilde{m}_j \frac{\tilde{z}_j - z_k}{|\tilde{z}_j - z_k|^3}, \tag{3.7}$$

and

$$\ddot{\tilde{z}}_k = \sum_j m_j \frac{z_j - \tilde{z}_k}{|z_j - \tilde{z}_k|^3} + \sum_{j \neq k} \tilde{m}_j \frac{\tilde{z}_j - \tilde{z}_k}{|\tilde{z}_j - \tilde{z}_k|^3}. \tag{3.8}$$

R. Moeckel and C. Simo [7] have proved the following result: If $\varphi = 0$, then for every mass ratio b, there are exactly two planar central configurations consisting of two nested regular N-gons. For one of these, the ratio of the sizes of the two polygons is less than 1, and for the other it is greater than 1.

Xie-Zhang [12,16] has studied the inverse problem of the theorem of Moeckel-Simo and the following results were established.

Theorem 3.1 If, for $N \geq 2$, the functions $z_k(t)$ and $\tilde{z}_k(t)$ given by (3.5) and (3.6) are solutions of the 2N-body problem (3.7) and (3.8), it follows that $\omega^2 = M\gamma/N$, where

$$\gamma = \frac{1}{1+b}\left(\sum_{j=1}^{N-1} \frac{1-\rho_j}{|1-\rho_j|^3} + \sum_{j=0}^{N-1} \frac{b(1-a\rho_j e^{i\varphi})}{|1-a\rho_j e^{i\varphi}|^3}\right). \tag{3.9}$$

Here, b and a have the following relationship:

$$b = \frac{\left(\sum_{j=0}^{N-1} \frac{ae^{i\varphi}-\rho_j}{|ae^{i\varphi}-\rho_j|^3} - \sum_{j=1}^{N-1} \frac{ae^{i\varphi}(1-\rho_j)}{|1-\rho_j|^3}\right)}{\left(\sum_{j=0}^{N-1} \frac{ae^{i\varphi}(1-a\rho_j e^{i\varphi})}{|1-a\rho_j e^{i\varphi}|^3} - \sum_{j=1}^{N-1} \frac{ae^{i\varphi}(1-\rho_j)}{a^3|1-\rho_j|^3}\right)}. \tag{3.10}$$

Theorem 3.2 For $N \geq 2$, $m_k > 0$, and $\tilde{m}_k = bm_k, b > 0$, the functions $z_k(t)$ and $\tilde{z}_k(t)$ given by (3.5) and (3.6) with $\omega^2 = M\gamma/N$ and γ given by (3.9) are solutions of the 2N-body problem (3.7) and (3.8) if and only if $m_1 = m_2 = \cdots = m_N$ and b is determined uniquely by (3.10).

Corollary 3.1 For $N = 2$, $\varphi = 0$ and $a > 1$ the functions $z_k(t)$ and $\tilde{z}_k(t)$ given by (3.5) and (3.6) with $\omega^2 = M\gamma/N$ and $\gamma = \frac{1}{1+b}(\frac{1}{4} - \frac{4ab}{(a^2-1)^2})$ are solutions of the 2×2-body problem (3.7) and (3.8) if and only if $m_1 = m_2$ and $b = \frac{a^7-2a^5-8a^4+a^3-8a^2}{17a^4-2a^2+1}$. This is a collinear periodic solution for the 4-body problem, when $0 < a < 1$, $\gamma = \frac{1}{1+b}(\frac{1}{4} + \frac{2b(a^2+1)}{(a^2-1)^2})$ and $b = \frac{a^7-2a^5+17a^3}{-8a^5+a^4-8a^3-2a^2+1}$.

Corollary 3.2

(i) Let $N = 2$ and $\varphi = \frac{\pi}{2}$. Then (3.10) is equivalent to

$$b = [(1+a^2)^{-3/2} - 2^{-3}] \cdot [(1+a^2)^{-3/2} - (2a)^{-3}]^{-1}. \tag{3.11}$$

(ii) For $N = 2$ and $\varphi = \frac{\pi}{2}$, $b = 1$ if and only if $a = 1$. That is $z_k and \tilde{z}_k$ are solutions of (3.7) and (3.8), $m_1 = m_2 = \tilde{m}_1 = \tilde{m}_2$ if and only if $m_1 = m_2 = \tilde{m}_1 = \tilde{m}_2$ are at the vertices of a square.

(iii) For $N = 2$ and $\varphi = \frac{\pi}{2}$, i.e., $\tilde{m}_1 = bm_1, \tilde{m}_2 = bm_2$ are at the vertices of a diamond. Then $m_1, m_2, \tilde{m}_1, \tilde{m}_2$ form a central configuration if and only if $\frac{\sqrt{3}}{3} < a < \sqrt{3}$, where a stands for the ratios of its diagonals, and the ratio b of masses is determined uniquely by (3.11), and $m_1 = m_2$.

Special cases are the degenerate central configurations where $a = \frac{\sqrt{3}}{3}$, $b = \infty$ and where $a = \sqrt{3}$, $b = 0$.

Corollary 3.3 Let $N \geq 2, m_k > 0$ and $\tilde{m}_k = bm_k, b > 0$. If $a = 1$ and $\varphi = \frac{\pi}{N}$, then $z_k(t)$ and $\tilde{z}_k(t)$ given by (3.5) and (3.6) with $\omega^2 = M\gamma/N$ and γ given by (3.9) are solutions of the 2N-body problem (3.7) and (3.8) if and only if $m_1 = m_2 = \cdots = m_N = \tilde{m}_1 = \cdots = \tilde{m}_N$.

4 Central Configurations in R^3

Definition 4.1 A central configuration of $N+1$ bodies, N of which are coplanar, the $(N+1)-th$ being off the plane, is called a pyramidal central configuration (p.c.c.). Equivalently, we will say that the c.c. has the shape of a pyramid. In 1904, Pizzetti proved that the regular tetrahedron is the unique non-flat c.c. corresponding to four arbitrary masses. His result is equivalent to the following theorem [9]:

Theorem (P. Pizzetti) For four arbitrary masses there exists only one non-flat c.c., which is the regular tetrahedron.

In 1996, Nelly Faycal established his classification of all p.c.c.'s with a rectangular base for the five-body problem. Her result is equivalent to the following theorem [4]:

Theorem (Nelly Faycal) A 5-body system with a rectangular base form a p.c.c. if and only if the base is a square and the masses of four bodies which lie at the vertices of the square base are equal.

Xie-Zhang [13] proved:

Theorem 4.1 Let $\tilde{q} = (\tilde{q}_1, \cdots, \tilde{q}_{N+1})(N \geq 3)$ be a p.c.c., such that $\tilde{q}_{N+1}$ is at the top vertex which is off the plane containing $m_1, \cdots, m_N$, then the base $m_1, \cdots, m_N$ with position $\tilde{q}_1, \cdots, \tilde{q}_N$ is also a central configuration (c.c.) with multiplier $\lambda_1 = \frac{M_N}{|q_{N+1}-q_1|^3}$, where $M_N = \sum_{i=1}^N m_i$.

Theorem 4.1′ If $m_1, m_2, \cdots, m_N$ with position $q_1, q_2, \cdots, q_N$, are coplanar and form a c.c. with multiplier λ_0 and if there exists a position q_{N+1} such that $\frac{1}{|q_{N+1}-q_i|^3} = \frac{\lambda_0}{M_N}(1 \leq i \leq N)$, where $M_N = \sum_{i=1}^N m_i$, then for any mass m_{N+1} with position q_{N+1}, $m_1, m_2, \cdots, m_{N+1}$ form a p.c.c.

Theorem 4.2 Let $\tilde{q} = (\tilde{q}_1, \cdots, \tilde{q}_5)$ be a p.c.c., such that $\tilde{q}_5$ is at the top vertex which is off the plane containing $m_1, \cdots, m_4$, then $\tilde{q}_1, \cdots, \tilde{q}_4$ are at the vertices of a square (its base is a square) if and only if $m_1 = m_2 = m_3 = m_4$.

Theorem 4.3 For $N+1(N \geq 2)$ body problems with masses $m_1, m_2, \cdots, m_{N+1} \in R^+$, and positions $\tilde{q}_1, \cdots, \tilde{q}_{N+1} \in R^3$, assume $\tilde{q}_1, \cdots, \tilde{q}_N$ are coplanar and lie at the vertices of a regular polygon inscribed on a unit circle, the $(N+1)-th$ is off the plane, and the $N+1$ bodies form a p.c.c., then

$$\frac{1}{D_{N+1,k}^3} = \frac{1}{4N}\sum_{j=1}^{N-1}\csc(\frac{\pi j}{N}), \qquad 1 \leq k \leq N, \tag{4.1}$$

where $D_{k,j} = |\tilde{q}_k - \tilde{q}_j|$. Note that there is no loss of generality in assuming that the regular polygon is inscribed on the unit circle since the c.c. (1.3) is invariant under the transformation $\tilde{q}_k \to \tilde{q}_k/a, \lambda \to a^2\lambda$.

Theorem 4.4 For $N+1(N \geq 4)$ body problems with masses $m_1, m_2, \cdots, m_{N+1} \in$

R^+, and positions $\tilde{q}_1, \cdots, \tilde{q}_{N+1} \in R^3$, assume $\tilde{q}_1, \cdots, \tilde{q}_N$ are coplanar and lie at the vertices of a regular polygon inscribed on a unit circle, the $(N+1)-th$ is off the plane, then the $N+1$ bodies form a p.c.c. if and only if $m_1 = m_2 = \cdots = m_N$ and (4.1) holds.

Definition 4.2 A central configuration of $N+2$ bodies, N of which are coplanar, the other two being off the plane and in opposite directions, is called a double pyramidal central configuration (d.p.c.c.).

Zhang-Zhou [17] proved :

Theorem 4.5 Let $\tilde{q}_1, \cdots, \tilde{q}_N, \tilde{q}_{N+1}, \tilde{q}_{N+2}$ be a d.p.c.c. such that $\tilde{q}_{N+1}$ and $\tilde{q}_{N+2}$ are at the end points of the vertical line of the plane Oxz containing $m_1, \cdots, m_N$, if $\sum_{i=1}^N m_i \tilde{q}_i = 0$ and $\tilde{q}_1, \cdots, \tilde{q}_N$ are on a circle and the vertical line passes the origin O, and $m_{N+1} = m_{N+2}$, then the base with position $\tilde{q}_1, \cdots, \tilde{q}_N$ is also a central configuration (c.c.).

Theorem 4.6 Under the assumptions of Theorem 4.5, if $\tilde{q}_1, \cdots, \tilde{q}_N$ lie at the vertices of a regular polygon inscribed on a unit circle, then we have

$$\frac{2m_{N+1}}{8d^3} + \frac{(\sum_{k=1}^N m_k)^2 - (2m_{N+1})^2}{M \cdot D^3} = (\sum_{k=1}^N m_k) \cdot \frac{1}{4N} \sum_{k=1}^{N-1} \csc(\frac{\pi k}{N}), \tag{4.2}$$

where

$$D = |\tilde{q}_{N+1} - \tilde{q}_i|, \qquad 1 \le i \le N, \tag{4.3}$$

$$d = \frac{1}{2} |\tilde{q}_{N+1} - \tilde{q}_{N+2}|. \tag{4.4}$$

Theorem 4.7 Under the assumptions of Theorem 4.6, the $N+2$ bodies form d.p.c.c. if and only if $m_1 = \cdots = m_N$ and (4.2) holds.

Theorem 4.8 Under the assumptions of Theorem 4.5, if $N = 4$ then $\tilde{q}_1, \cdots, \tilde{q}_4$ are at the vertices of a square (its base is a square) if and only if $m_1 = m_2 = m_3 = m_4$.

Without special assumptions, $m_1, m_2, \cdots, m_N$ are always coplanar, m_{N+1} lies upward the plane and m_{N+2} lies downward the plane. For convenience of expression, some notations are taken such as

$$M_N = \sum_{i=1}^N m_i, M_{N+1} = \sum_{i=1}^{N+1} m_i, M_{N+2} = \sum_{i=1}^{N+2} m_i, \tag{4.5}$$

$$M'_{N+1} = \sum_{i=1, i \ne N+1}^{N+2} m_i, D_{i,j} = |q_i - q_j|. \tag{4.6}$$

Xie-Zhang [14] proved

Theorem 4.9 If $m_1, m_2, \cdots, m_N, m_{N+1}$ form a p.c.c. with multiplier $\lambda_1 = \frac{M_{N+1}}{D^3_{N+1,i}}$ and $m_1, m_2, \cdots, m_N, m_{N+1}, m_{N+2}$ form a d.p.c.c., then $m_1, m_2, \cdots, m_N, m_{N+2}$ form a p.c.c. and $D_{N+1,i} = D_{N+2,j} = D_{N+2,N+1}, 1 \le i, j \le N$.

Theorem 4.10: If $m_1, m_2, \cdots, m_N, m_{N+1}$ form a p.c.c. with multiplier $\lambda_1 = \frac{M_{N+1}}{D^3}$

and $m_1, m_2, \cdots, m_N, m_{N+2}$ form a p.c.c. with multiplier $\lambda_3 = \frac{M'_{N+1}}{D^3}$, then $D_{N+1,i} = D_{N+2,j}$, $(1 \le i, j \le N)$ (obviously). And if

$$D_{N+1,i} = D_{N+2,j} = D_{N+2,N+1} = D, (1 \le i, j \le N)$$

then $m_1, m_2, \cdots, m_N, m_{N+1}, m_{N+2}$ form a d.p.c.c.

Theorem 4.11 If $m_1, m_2, \cdots, m_N, m_{N+1}, m_{N+2}$ form a d.p.c.c. then
(i) If $D_{N+1,i} = D_{N+2,j} = D_{N+2,N+1} = D, (1 \le i, j \le N)$, then $m_1, m_2, \cdots, m_N, m_{N+1}$ form a p.c.c.; $m_1, m_2, \cdots, m_N, m_{N+2}$ form a p.c.c. and $m_1, m_2, \cdots, m_N$ form a c.c.
(ii) If $D_{N+1,i} = D_{N+2,j} = D, (1 \le i, j \le N)$ and if $\frac{1}{M_{N+2}} \sum_{j=1}^{N+2} m_j q_j = 0$, $\frac{1}{M_N} \sum_{j=1}^{N} m_j q_j = 0$, then $m_1, \cdots, m_N$ form a c.c.

Theorem 4.12 If the base is a regular polygon inscribed on a unit circle, and $N+1$ bodies form a p.c.c, then the $N+2$ bodies can't form a d.p.c.c satisfying Theorem 4.9.

In the following,we consider the configurations in R^3 consisting of two layer regular N-gons ($N \ge 2$). We assume that the lower layer regular N-gon lies in the horizontal plane, and the upper layer regular N-gon parallels the lower one and the z-axis passes through both the centers of two regular N-gons. Suppose that the lower layer particles have masses $m_1, \cdots, m_N$ and the upper layer particles have masses $\tilde{m}_1, \cdots, \tilde{m}_N$ respectively. Then these assumptions can be interpreted as follows. Let ρ_k denote the N complex $k-th$ roots of unity, i.e.,

$$\rho_k = \exp(\frac{2\pi k}{N} i), \tag{4.7}$$

$$\tilde{\rho}_k = a\rho_k \cdot e^{i\varphi}, \tag{4.8}$$

where

$$a > 0, 0 \le \varphi \le 2\pi. \tag{4.9}$$

We assume that $m_k (1 \le k \le N)$ located at the vertices ρ_k of the lower layer regular N-gons. Let

$$q_k = (\rho_k, 0), \tag{4.10}$$

$$\tilde{q}_k = (\tilde{\rho}_k, h), \qquad (h > 0), \tag{4.11}$$

where h is the distance between the two layers. Then the center of the masses is

$$z_0 = \sum_j (m_j q_j + \tilde{m}_j \tilde{q}_j)/M, \tag{4.12}$$

where

$$M = \sum_j (m_j + \tilde{m}_j). \tag{4.13}$$

Let

$$P_k = q_k - z_0, \tag{4.14}$$

$$\tilde{P}_k = \tilde{q}_k - z_0. \tag{4.15}$$

If $m_1, \cdots, m_N; \tilde{m}_1, \cdots, \tilde{m}_N$ form a c.c., then there is a $\lambda \in R$ such that

$$\sum_{j \neq k} m_j \frac{P_j - P_k}{|P_j - P_k|^3} + \sum_j \tilde{m}_j \frac{\tilde{P}_j - P_k}{|\tilde{P}_j - P_k|^3} = -\lambda P_k, \quad 1 \leq k \leq N, \tag{4.16}$$

and

$$\sum_j m_j \frac{P_j - \tilde{P}_k}{|P_j - \tilde{P}_k|^3} + \sum_{j \neq k} \tilde{m}_j \frac{\tilde{P}_j - \tilde{P}_k}{|\tilde{P}_j - \tilde{P}_k|^3} = -\lambda \tilde{P}_k, \quad 1 \leq k \leq N. \tag{4.17}$$

In the equations (4.16) and (4.17) and in the following, unless other restricted, all indices and summations will range from 1 to N.
Under the case $\varphi = 0$, Mockel-Simo [7] proved
Theorem (Mockel-Simo) If $N < 473$, there is a unique pair of spatial central configurations of parallel regular N-gons. If $N \geq 473$, there are no such central configurations for $b < b_0(N)$,where b is the mass ratio. At $b = b_0$ a unique pair bifurcations from the planar central configuration with the smaller masses on the inner polygon. This remains the unique pair of spatial central configurations until $b = 1/b_0$, where a similar bifurcation occurs in reverse, so that for $b > 1/b_0$, only the planar central configurations remain.
Xie-Zhang-Zhou [15] proved that
Theorem 4.13 Under the above assumptions, if $m_1, \cdots, m_N; \tilde{m}_1, \cdots, \tilde{m}_N$ form a c.c. then (i) $\lambda = \frac{M}{N} \sum_j \frac{1}{|q_j - \tilde{q}_N|^3}$, and
(ii) For $N \geq 2, (\varphi \neq \frac{\pi}{2}$, when $N = 2)$ $m_1 = m_2 = \cdots = m_N$ and $\tilde{m}_1 = \tilde{m}_2 = \cdots = \tilde{m}_N$.
Theorem 4.14 Under the above assumptions and $\tilde{m}_j = bm_j$, then $m_1, \cdots, m_N$; $\tilde{m}_1, \cdots, \tilde{m}_N$ form a c.c. if and only if the parameters b, a, h and φ satisfy the following relationships

$$\frac{\lambda}{M} = \frac{1}{N(1+b)} \left(\sum_{j \neq N} \frac{1 - \rho_j}{|q_j - q_N|^3} + \sum_j \frac{b(1 - a\rho_j e^{i\varphi})}{|\tilde{q}_j - q_N|^3} \right), \tag{4.18}$$

$$\frac{\lambda}{M} = \frac{1}{N} \sum_j \frac{1}{|\tilde{q}_j - q_N|^3}, \tag{4.19}$$

$$\frac{\lambda}{M} = \frac{e^{-i\varphi}}{Na(1+b)} \left(\sum_j \frac{ae^{i\varphi} - \rho_j}{|q_j - \tilde{q}_N|^3} + \sum_{j \neq N} \frac{b(a - a\rho_j)e^{i\varphi}}{|\tilde{q}_j - \tilde{q}_N|^3} \right). \tag{4.20}$$

Corollary 4.1 i) When $N = 2$ and $\varphi = 0$, if $m_1, m_2, \tilde{m}_1, \tilde{m}_2$ form a c.c. then the parameters a,b,h satisfy

$$b = \frac{1 - (x + y)}{4a(y - x)}, \tag{4.21}$$

and

$$\frac{4a^2(x-y)}{4a^3(x+y)-1} = \frac{1-(x+y)}{4a(y-x)}, \tag{4.22}$$

where $x = \frac{1}{(\sqrt{(1+a)^2+h^2})^3}, y = \frac{1}{(\sqrt{(1-a)^2+h^2})^3}$. These are the conditions that a regular trapezoid forms a c.c.

ii) $a = 1, h = 2$ while $b = 1$, which implies that if four equal masses form a c.c., then they lie at the vertices of a square .

Corollary 4.2 When $N = 4$, $\varphi = 0$ and $b = 1$, if $m_1, \cdots, m_4, \tilde{m}_1, \cdots, \tilde{m}_4$ form a c.c. then $a = 1$, $h = \sqrt{2}$ are a solution, i.e., the eight bodies with equal masses lying at the vertices of a regular hexahedron form a c.c.

Theorem 4.15 Under the above assumptions, when $N = 2$ and $\varphi = \frac{\pi}{2}$, then $m_1, m_2, \tilde{m}_1, \tilde{m}_2$ form a c.c. if and only if $a = 1, h = \sqrt{2}$. It indicates that if four bodies form a c.c. in R^3 then four bodies lie at the vertices of a tetrahedron.

Acknowledgement

The author would like to thank his graduate student Xie Zhifu for some discussions.

References

[1] A. Albouy, The symmetric central configuration of four equal masses, Contemp. Math. 198 (1996) 131-135.

[2] F. Diacu and P. Holmes, Celestial Encounters: The Origins of Chaos and Stability, Princeton University Press, 1996.

[3] B. Elmabsout, Sur l'existence des certaines configurations d'equilibre relatif dans le probleme des N corps, Celest. Mech. 41 (1988) 131-151.

[4] N. Faycal, On the classification of pyramidal central configurations, Proc. AMS. 124 (1996) 249-258.

[5] J. Lagrange, Oeuvres, Vol. 6, Gauthier-Villars, Paris, 1873.

[6] M. Marcus and H. Minc, A survey of matrix theory and matrix inequalities, Allyn and Bacon, Boston, Mass, 1964.

[7] R. Mockel and C. Simo, Bifurcation of spatial central configurations from planar ones, SIAM J. Math. Anal. 26 (1995) 978-998.

[8] L. M. Perko and E. L. Walter, Regular polygon solutions of the N-body problem, Proc. AMS, 94 (1985) 301-309.

[10] A. Wintner, Analytical foundations of celestial mechanics, Princeton Univ. Press, 1941.

[11] X. Zhifu and Z. Shiqing, A simpler proof of regular polygon solutions of the N-body problem, Physics Letters A 277 (2000) 156-158.

[12] X. Zhifu and Z. Shiqing, Nested regular polygon solutions of 2N-body problems, submitted.

[13] X. Zhifu and Z. Shiqing, On the pyramidal central configurations with regular polygonal base, preprint.

[14] X. Zhifu and Z. Shiqing, Double pyramidal central configurations for N+2-body problems, preprint.

[15] X. Zhifu, Z. Shiqing and Z. Qing, Two layer regular polygon central configurations in R^3, preprint.

[16] X. Zhifu and Z. Shiqing, On central configurations for planar 2N-body problems, preprint.

[17] Z. Shiqing and Z. Qing, Double pyramidal central configurations, preprint.

Theoretical Physics 2002, Part 1
Edited by Thomas F. George and Henk F. Arnoldus, pp. 133-179
ISBN 1-59033-435-3

Geometry and Topology of Solitons

N. Riazi*

Abstract

Basic concepts and definitions in differential geometry and topology which are important in the theory of solitons and instantons are reviewed. Many examples from soliton theory are discussed briefly, in order to highlight the application of various geometrical concepts and techniques.

1 Introduction

Recent developments in soliton theory have been associated with frequent application of geometrical and topological ideas which provide an elegant interpretation of many soliton properties. The mathematical methods of differential geometry and topology are very abstract and rigorous which make them hard to grasp by a physics-oriented scientist. In this writing, I have tried to provide an interdisciplinary and informal introduction to those topics in differential geometry and topology which have proven important in soliton theory. A background knowledge of tensor calculus and field theory is assumed throughout this manuscript. In the following sections, we have made extensive use of Eguchi et al. (1980) and Drechsler and Mayer (1977).

2 Forms, Fibers, and Bundles

An $n-$dimensional *manifold* is a space which behaves locally like R^n. In a similar manner, complex manifolds can be defined which are locally similar to C^n. A circle is a simple example of a one-dimensional manifold while a figure like + is not a manifold because of its behavior at the junction point. *Compact manifolds* have a finite volume. An immediate example is the $n-$dimensional sphere S^n (S^o which contains only two points can be included in this definition). In contrast, R^n is an example of a noncompact manifold.

Group manifolds are spaces constructed by the free parameters which specify the elements of a group. For example, the group manifold of Z_2 is S^o while that of $U(1)$ is S^1, that of $SU(2)$ is S^3, that of $SO(3)$ is $SU(2)/Z_2$ or $P_3(R)$, etc.

*Department of Physics, Shiraz University, Shiraz 71454, Iran and Department of Physics and Astronomy, University of Victoria, Victoria, V8P 1A1, Canada

Consider a real vector space E on the $n-$dimensional manifold $\mathcal{M}$ (we are mainly concerned with R^n and the Minkowski spacetime). *1-forms* are linear mappings from E to R:

$$\omega(\alpha u + \beta v) = \alpha\omega(u) + \beta\omega(v), \tag{1}$$

where ω is a 1-form , $\alpha, \beta \in R$, and $u, v \in E$. A familiar 1-form in classical mechanics is the work 1-form $\vec{F}.d\vec{x}$. *p−forms* $\omega(u_1, ..., u_p)$ are p-linear antisymmetric mappings from p-vectors $E \times ... \times E$ (p-times) to R:

$$\omega(u_1, ..., \alpha u'_j + \beta u''_j, ..., u_p) = \alpha\omega(u_1, ..., u'_j, ..., u_p) + \beta\omega(u_1, ..., u''_j, ..., u_p), \tag{2}$$

with

$$\omega(u_{i_1}, ..., u_{i_p}) = sgn(\pi)\omega(u_1, ..., u_p), \tag{3}$$

where $sgn(\pi)$ =+1(-1) if the permutation $\pi : (1, ..., p) \rightarrow (i_1, ..., i_p)$ is even (odd). Consider the $p-$form ω and the $q-$form η such that $p + q$ is less than or equal to the dimension of the vector space E. The *exterior or wedge product* $\omega \wedge \eta$ is a $(p + q)-$form such that

$$\eta \wedge \omega = (-1)^{pq}\omega \wedge \eta. \tag{4}$$

This product is distributive with respect to addition, and associative:

$$\omega \wedge (\eta + \zeta) = \omega \wedge \eta + \omega \wedge \zeta,$$

and

$$\omega \wedge (\eta \wedge \zeta) = (\omega \wedge \eta) \wedge \zeta. \tag{5}$$

The space of all tangent vectors to a manifold $\mathcal{M}$ at the point x is called the *tangent space* and is denoted by $T_x\mathcal{M}$. This space has the same dimension n as the manifold $\mathcal{M}$. The union of all such tangent spaces form the *tangent bundle* $T\mathcal{M}$ and is a manifold of dimension $2n$.

While $\{\frac{\partial}{\partial x^i}\}$ forms a basis of the tangent space $T_x\mathcal{M}$, the dual basis $\{dx^i\}$ forms a basis of the so-called *cotangent space* $T^*_x(\mathcal{M})$. The inner product of these two bases satisfy

$$(\frac{\partial}{\partial x^i}|dx^j) = \delta^j_i. \tag{6}$$

Differential p−forms are $p-$forms on the tangent space $T\mathcal{M}$. Consider, for example, a real function $F(x, y, z)$ on the three dimensional Euclidean space R^3. F is in fact a 0-form, while $dF = \partial_i F dx^i$ is a differential 1-form, similar to the work form. The flux 2-form

$$\Phi = \Phi_x dy \wedge dz + \Phi_y dz \wedge dx + \Phi_z dx \wedge dy \tag{7}$$

is another example of a differential form. In Minkowski spacetime, the electromagnetic potentials form a 1-form A with components A_μ ($\mu = 0, 1, 2, 3$). The components of a $p-$form coincide with those of an anti-symmetric covariant tensor of rank p for $p > 1$. For $p = 0$, the 0-form transforms like a scalar and $p = 1$ forms correspond to covariant vectors.

The tensor product $\omega \otimes \eta$ of the $p-$form ω with the $q-$form η does not make a $(p+q)-$form since $\omega \otimes \eta$ is not antisymmetric with respect to all of its components. The wedge product $\omega \wedge \eta$ is defined in such a way to preserve the antisymmetry. For example, the wedge product of two 1-forms η and ω satisfies $\omega \wedge \eta = \omega \otimes \eta - \eta \otimes \omega$. 1-forms, therefore, anticommute ($\omega \wedge \eta = -\eta \wedge \omega$).

The exterior derivative of a $p-$form ω is defined in such a way to lead to a $(p+1)$-form. Consider, for example, the 1-form A on M:

$$A = A_o dt + A_1 dx + A_2 dy + A_3 dz = A_\mu dx^\mu. \tag{8}$$

We have

$$dA = dA_o \wedge dt + dA_1 \wedge dx + dA_2 \wedge dy + dA_3 \wedge dz = dA_\mu \wedge dx^\mu$$

$$= (\partial_\alpha A_\mu dx^\alpha) \wedge dx^\mu = \partial_\alpha A_\mu dx^\alpha \wedge dx^\mu$$

$$= (\partial_y A_3 - \partial_z A_2) dy \wedge dz + ... \tag{9}$$

The *Hodge operation* * on a $p-$form ω produces an $(n-p)$-form $^*\omega$ according to

$$(*\omega)_{ij...} = \frac{1}{p!} \epsilon_{lm...ij...} \omega^{lm...}, \tag{10}$$

where $\epsilon_{lm...}$ (n indices) is the totally antisymmetric tensor of the $n-$dimensional space. Note that * is defined in terms of components (the components of $p-$forms have p indices). The antisymmetry of differential forms implies

$$dd\omega = 0 \tag{11}$$

for any $p-$form ω. It can also be shown that

$$d(\omega \wedge \eta) = d\omega \wedge \eta + (-1)^p \omega \wedge d\eta \tag{12}$$

for any $p-$form ω and $q-$form η. The $p-$form ω is called *closed* (globally) if $d\omega = 0$ and *exact* if $\omega = d\eta$ where η is a $(p-1)$-form. Note that all exact forms are closed, but the converse is not always true. The reader recalls that not all vector fields can be written as the gradients of scalar functions.

The electromagnetic field 2-form F defined by

$$F = dA = \frac{1}{2}F_{\mu\nu}dx^{\mu} \wedge dx^{\nu} \tag{13}$$

is an example of an exact form. The action for the free Maxwell field can be written in the form

$$-\int \frac{1}{2} \; {}^{*}dA \wedge dA. \tag{14}$$

Note that if $F_{\alpha\beta}$ are the components of a two-form F, $(dF)_{\alpha\beta\gamma} = \partial_{\alpha}F_{\beta\gamma} + \partial_{\beta}F_{\gamma\alpha} + \partial_{\gamma}F_{\alpha\beta}$, which is antisymmetric with respect to all indices. It can also be shown that if ω is closed and η is exact, $\omega \wedge \eta$ will be exact. The reader can verify these two statements as exercises.

The well-known identities $\vec{\nabla} \times \vec{\nabla}\phi = 0$ and $\vec{\nabla}.\vec{\nabla} \times \vec{F} = 0$ in vector analysis follow from the generalized identity $dd = 0$. Consider, for example, the 1-form

$$\omega = u_i dx^i \tag{15}$$

in the three dimensional Euclidean space where u_i ($i = 1, 2, 3$) are functions of x^i ($x^1 = x$, $x^2 = y$, $x^3 = z$). The exterior derivative of this 1-form yields $\vec{\nabla} \times \vec{u}$:

$$d\omega = d(u_i dx^i) = \frac{\partial u_i}{\partial x^j} dx^j \wedge dx^i,$$

$$= \frac{1}{2}(\frac{\partial u_i}{\partial x^j} - \frac{\partial u_j}{\partial x^i})dx^j \wedge dx^i = \frac{1}{2}\epsilon_{kji}(\vec{\nabla} \times \vec{u})_k dx^j \wedge dx^i, \tag{16}$$

and

$$dd\omega = \frac{1}{2}(\frac{\partial^2 u_i}{\partial x^k \partial x^j} - \frac{\partial^2 u_j}{\partial x^k \partial x^i})dx^k \wedge dx^j \wedge dx^i = 0. \tag{17}$$

The last identity follows from the symmetry of $\frac{\partial^2}{\partial x^k \partial x^j}$ and $\frac{\partial^2}{\partial x^k \partial x^i}$ and antisymmetry of $dx^k \wedge dx^j$ and $dx^k \wedge dx^i$.

A $p-$form which can be expressed as $\omega = d\phi_1 \wedge ... \wedge d\phi_p$ where $\phi_1, ...\phi_p$ are scalar fields, is called *simple*. A simple form is both exact and closed. Note that ω can be written as $\omega = d(\phi_1 d\phi_2 \wedge ... \wedge d\phi_p)$.

The *co-differential* of the $p-$form ω is defined as

$$\delta\omega = (-1)^{p(n-p+1)} \; {}^{*}d^{*}\omega \tag{18}$$

for Euclidean metrics. The definition for Minkowskian metrics differ by a - sign. Note that if ω is a $p-$form, $\delta\omega$ is a $(p-1)-$form. Also $\delta\delta = \pm \; {}^{*}d^{2*} = 0$. The Laplacian operator is obtained by forming $\Delta = -(\delta d + d\delta)$, which in a Minkowski spacetime becomes the d'Alembertian operator. A differential form ω is called *harmonic* if $\Delta\omega = 0$. The inhomogeneous Maxwell equations can also be written in the form $\delta F = J$ and the conservation of electric current as $\delta J = 0$.

Integration of $p-$forms over the space of interest (or part of it) is of great importance. In R^n, $dx^1 \wedge ... \wedge dx^n = dx^1 ... dx^n$ provides the volume $n-$form. For a general metric, the volume form is given by $\sqrt{|g|}dx^1...dx^n$, where g is the determinant of the metric tensor. *Stokes's theorem* reads

$$\int_V d\omega = \int_{\partial V} \omega, \tag{19}$$

where the LHS integration is over the submanifold V with the boundary ∂V. The manifold over which the integration is performed (and hence its boundary) is assumed to be *orientable.* The existence of a volume form on the manifold of interest guarantees its orientability. *Klein bottle* and *Möbius strip* are examples of nonorientable manifolds. Stokes's theorem leads to the more common Stokes and divergence theorems in vector analysis.

Consider a 3-dimensional spacelike volume V with the boundary ∂V. The magnetic flux through ∂V is given by $\int_{\partial V} F$, while the electric flux is given by $\int_{\partial V} {}^*F$. The electric charge contained in V is given by $-\int_V {}^*J$. The Gauss theorem can therefore be written as

$$\int_{\partial V} {}^*F = -\int_V {}^*J.$$

The following results are relevant to our discussion:

- Any $k-$dimensional regular domain X of a manifold $\mathcal{M}$ has a boundary ∂X which is a $(k-1)$-dimensional compact manifold. ∂X itself has no boundary, i.e. $\partial\partial X = 0$. This is called *Cartan's lemma* .
- If $\mathcal{M}$ is a simply connected manifold (see section 9), then $\oint_C \omega = 0$ for an arbitrary closed 1-form ω. C is a closed curve in $\mathcal{M}$.
- If $\mathcal{M}$ is simply connected, then all closed 1-forms ω on $\mathcal{M}$ are exact.
- Consider a compact, oriented, $p-$dimensional submanifold X of the simply connected manifold $\mathcal{M}$. Then, for any two cohomologous (see section 8) $p-$forms ω and η on $\mathcal{M}$, $\int_X \omega = \int_X \eta$.
- A closed $p-$form ω on S^p is exact if and only if $\int_{S^p} \omega = 0$.

Note that any singularities in the fields described by differential forms must be treated as holes in the manifold on which the forms are defined. For example, consider $d\theta$ on the xy-plane where $\theta = arctan(y/x)$:

$$d\theta = \frac{-y}{x^2+y^2}dx + \frac{x}{x^2+y^2}dy. \tag{20}$$

This form is closed but not exact. While every closed $p-$form ($p > 0$) on R^n is exact, (20) is not exact because of the singularity at $x = y = 0$ which introduces a hole in R^2. In other words, the manifold on which (20) is defined is $R^2 - \{0\}$. Any closed 1-form ω on $R^2 - \{0\}$ integrates on a smooth closed curve C according to

$$\oint_C \omega = w(C,0) \int_{S^1} \omega, \tag{21}$$

where $w(C, 0)$ is the winding number of C with respect to the origin (see section 3).

The *Hilbert product* of forms is defined as follows

$$< \omega|\eta >= \int_{\mathcal{M}} {}^*\omega \wedge \eta = \int_{\mathcal{M}} \frac{1}{p!}\omega^{i_1...i_p}\eta_{i_1...i_p}\sqrt{|g|}dx^1...dx^n, \tag{22}$$

where g is the determinant of the manifold's metric. Note that ω and η are both k-forms. For compact manifolds the Hilbert product is always well-defined. It can be easily shown that $< {}^*\omega| {}^*\eta >= \pm < \omega|\eta >$ depending on the signature being Euclidean (+) or Minkowskian (-).

As an example, consider the action for a real Klein-Gordon field

$$S = \int (\frac{1}{2}\partial^\mu\phi\partial_\mu\phi - \frac{1}{2}m^2\phi^2)\sqrt{|g|}d^4x,$$

which can be written as

$$S = \frac{1}{2} < d\phi|d\phi > -\frac{1}{2}m^2 < \phi|\phi >= \int (\frac{1}{2}\, {}^*d\phi \wedge d\phi - \frac{1}{2}m^2\, {}^*\phi \wedge \phi). \tag{23}$$

The tangent space $T_x\mathcal{M}$ is the prototype of a *fiber*. In order to define a fiber more generally, consider the triplet (T, B, π), where T is the *bundle space* (or *total space*), B is the base space and π is a C^∞ (i.e. infinitely differentiable) mapping from T to B. Such a triplet is called a *fibration*. An example is $(\mathcal{M}\times V, \mathcal{M}, \pi)$ where $\mathcal{M}\times V$ is the product manifold of $\mathcal{M}$ with an arbitrary $m-$dimensional vector space V. This is called a *local vector bundle of rank* m. The projection map π is simply $\pi(x, u) = x$ in this case, where $x \in \mathcal{M}$ and $u \in V$. Part of the total space which sits on top of the point $x \in B$ is $\pi^{-1}(x)$ and is called a *fiber*. The manifold T can therefore be considered as a collection of fibers or a *fiber bundle*. A C^∞ mapping $f : B \to T$ such that $\pi \circ f = Id$ is called a *section* of the fibration. Here, Id_B is the identity map on B ($Id_B(x) = x$).

A fiber bundle is *locally trivializable* if it can be described by the product manifold $U_i \times F$ where U_i is a neighborhood of the base manifold and F is the fiber. Since local properties are not sufficient to describe the global topology of the bundle, a set of *transition functions* ϕ_{ij} are defined which specify how the fibers are related to each other in the overlapping region of the two neighborhoods U_i and U_j. The transition function ϕ_{ij} is therefore defined as the mapping of the fibers on U_i to the fibers on U_j over the region $U_i \cap U_j$. For a trivial fiber bundle, all the transition functions can be the identity map. As we said before, $\{\frac{\partial}{\partial x^\mu}\}$ is the standard basis for the local frames of the tangent bundle $T\mathcal{M}$ while $\{dx^\mu\}$ is the basis for the cotangent bundle $T^*\mathcal{M}$. Since $\frac{\partial}{\partial x^\mu} = \frac{\partial x'^\alpha}{\partial x^\mu}\frac{\partial}{\partial x'^\alpha}$ and $dx'^\mu = \frac{\partial x'^\mu}{\partial x^\alpha}dx^\alpha$, they define transition functions between two overlapping neighborhoods U and U' of the tangent and cotangent bundles, respectively.

In gauge field theories, we encounter a special type of fibration which is called a *principal fibration.* A principal fibration consists of a base manifold $\mathcal{M}$, the structure group G, and the manifold P on which the Lie group G acts. Like in an ordinary fibration, π is a mapping from P to $\mathcal{M}$. The principal fibration, therefore, is denoted by $(P, G, \mathcal{M}, \pi)$. If G is a gauge group, a section in the principal fibration corresponds to the choice of a particular gauge. In the case that there exists a global C^∞ section, the principal fibration is said to be *trivializable* . A trivializable fibration is isomorphic to $(\mathcal{M} \times G, G, \mathcal{M}, pr_1)$.

Any fiber bundle is trivial if the base space is contractible. Non-trivial fiber bundles can therefore exist over topologically non-trivial base manifolds (like R^3-0, S^1, etc.). Only for a trivial principal bundle can one find a single gauge potential which is smooth over the entire base manifold.

The fiber of a vector bundle is a linear vector space. The transition functions of a vector bundle are elements of the *general linear group* of the vector space. Similarly, the transition functions of a principal bundle are elements of G acting by left multiplication. The associated vector bundle $PX_\rho V$ is defined using the representation $R(G)$ acting on the finite-dimensional vector space V. The *associated vector bundle* is based on the equivalence relation

$$(p, \rho(g) \circ v) \simeq (p \circ g, v) \quad \forall \quad p \in P, \quad v \in V, \quad g \in G, \quad \textit{and} \quad \rho \in R(G). \tag{24}$$

The tangent space to the bundle space T_pP can be decomposed into *vertical* and *horizontal* parts:

$$T_pP = H_p \oplus V_p. \tag{25}$$

The vertical part which corresponds to the action of G maps into a single point of the base space. The horizontal part leads to the *connection* 1-form which will be defined in section 4.

The space of spinors is also a vector space and *spinor bundles* can be constructed in a similar way as the vector bundles. The corresponding principal bundle has the spin group with the *Clifford algebra* as the tangent space at the identity element.

3 Smooth Maps and Winding Numbers

Consider a smooth map from the $n-$dimensional differentiable manifold $\mathcal{M}$ to the $m-$dimensional differentiable manifold $\mathcal{M}'$

$$f : \mathcal{M} \to \mathcal{M}'. \tag{26}$$

This mapping is called

- *surjective*, if $f(\mathcal{M}) = \mathcal{M}'$. In other words, for all $y \in \mathcal{M}'$ there is at least one element $x \in \mathcal{M}$ such that $f(x) = y$.

- *injective*, if for all $x_1, x_2 \in \mathcal{M}$ where $x_1 \neq x_2$, we have $f(x_1) \neq f(x_2)$. In other words, distinct points in $\mathcal{M}$ have distinct images.

- *bijective*, if it is both surjective and injective. The existence of a bijective mapping from $\mathcal{M}$ to $\mathcal{M}'$ ensures that the points in these two spaces are in one-to-one correspondence.

For details, the reader is referred to Long (1971).

Let x^i ($i = 1, ..., n$) and y^j ($j = 1, ..., m$) be coordinate systems in $\mathcal{M}$ and $\mathcal{M}'$, respectively. These coordinates are related by the map f. We can expand $y^j(x^i)$ around a point p in $\mathcal{M}$ with coordinates x^i_o

$$y^j(x^i_o + \Delta x^i) = y^j(x^i_o) + \left(\frac{\partial y^j}{\partial x^i}\right)_p \Delta x^i + ... \tag{27}$$

Note that $[\frac{\partial y^j}{\partial x^i}]$ is an $m \times n$ matrix called the *Jacobi matrix.* This matrix defines the appropriate linear mapping from $T_p(\mathcal{M})$ to $T_{f(p)}(\mathcal{M}')$ (see Felsager, 1983).

A well-defined transformation between the two tangent spaces requires the mapping to be a *diffeomorphism*, which in simple terms means that the map is bijective with an inverse f^{-1} which is smooth. For $m = n$, the Jacobi matrix becomes a non-singular square matrix. If $m < n$ and f is everywhere regular, the map is called an *immersion*, and $f(\mathcal{M})$ is a submanifold of $\mathcal{M}'$. An *embedding* is an immersion which is further required to be a *homeomorphism.* If $m > n$, and f is everywhere regular, it is called a *submersion.* We now assume that $\mathcal{M}$ and $\mathcal{M}'$ are both compact and orientable, and have the same dimension n. The *Brouwer degree* of the map f is defined as

$$deg(f) = \sum_{p_i} sgn|\frac{\partial y^j}{\partial x^i}|_{p_i}, \tag{28}$$

where $f(p_i)$ are the regular values in $\mathcal{M}'$. A point $x \in \mathcal{M}$ is said to be *regular* if $f'(x) \neq 0$, otherwise, it is a *critical point.* A point $\xi \in \mathcal{M}'$ is either the image of a regular point, the image of a critical point, or it is not the image of any point. To see what a regular value means, consider a map $f : S^1 \rightarrow S^1$. In this example, the map $\xi = \theta$ is of degree $+1$, $\xi = -\theta$ is of degree -1, and the map $\xi = \frac{1}{\pi}(\theta - \pi)^2$ has a vanishing degree. Note also that $\xi = n\theta$ is of degree n.

The Brouwer degree of a map measures the net number of times $\mathcal{M}'$ is covered when all the points in $\mathcal{M}$ are swept once. The integer $deg(f)$ is also called *winding number.* The Brouwer degree vanishes for maps which are not surjective.

Brouwer's theorem states that

$$\int_{\mathcal{M}} f^*\omega = deg(f) \int_{\mathcal{M}'} \omega, \tag{29}$$

where ω is a $p-$form on $\mathcal{M}'$ and $f^*\omega$ is its pullback, which is a $p-$form on $\mathcal{M}$ and will be defined in the next section.

4 Gauge Fields as Connections on principal bundles

Consider a gauge group represented by $m \times m$ complex matrices acting on a complex $m-$dimensional vector space V. A vector bundle can be constructed with fibers isomorphic to V. A connection on this vector bundle is a 1-form on the base space with values in $C(m \times m)$ (the space of complex $m \times m$ matrices). The connection 1-forms of $SU(N)$, for example, are anti-hermitian, traceless $N \times N$ matrices and at the same time 1-forms on the base manifold (e.g. the Minkowski space). For a smooth section S of the principal bundle, the covariant derivative is defined according to[1] (Drechsler and Mayer, 1977):

$$DS = dS + A \wedge S, \tag{30}$$

where $A = A_\mu dx^\mu$ is the connection 1-form. Note that each A_μ is a complex $N \times N$ matrix, expandable in terms of the bases of $SU(N)$:

$$A_\mu = A^a_\mu T_a. \tag{31}$$

In the case of $SU(2)$, $T_a = \tau_a$ (a=1,2,3) are the Pauli matrices

$$\tau_1 = \begin{pmatrix} i & 0 \\ 0 & -i \end{pmatrix}, \quad \tau_2 = \begin{pmatrix} 0 & 1 \\ -1 & 0 \end{pmatrix}, \quad \tau_3 = \begin{pmatrix} 0 & i \\ i & 0 \end{pmatrix}. \tag{32}$$

The Lie algebra of a group G is the tangent space to the group manifold at the identity element T_eG. The basis $\{T_a\}$ of this space obeys the algebra

$$[T_a, T_b] = f_{abc} T_c, \tag{33}$$

where f_{abc} are the *structure constants* of the group.

Unlike the exterior derivative of ordinary 1-forms for which we have $ddS = 0$, DDS does not always vanish. This quantity which is a 2-form leads in a natural way to the concept of curvature. Taking the covariant derivative of (31),

$$DDS = D(A \wedge S) = dA \wedge S + A \wedge DS = F \wedge S, \tag{34}$$

[1] More commonly, covariant derivative is defined as $DS = dS + iA \wedge S$ such that τ_is become hermitian matrices.

where the curvature 2-form is

$$F = dA + A \wedge A. \tag{35}$$

It can be shown that the covariant derivative of F vanishes

$$DF = dF + A \wedge F - F \wedge A = 0, \tag{36}$$

which is the generalized form of the *Bianchi identity.* Like A, F is also in the form of an $m \times m$ complex matrix. F is called curvature because it is related to the Gaussian curvature when a curved manifold is concerned. The Bianchi identity constitutes one of the basic equations governing the gauge field A. An element g of the gauge group linearly transforms a vector $v \in V$

$$v \rightarrow gv. \tag{37}$$

Such a transformation is associated with the following gauge transformations on the connection and curvature:

$$A \rightarrow Ad(g^{-1})A + g^{-1}dg, \tag{38}$$

and

$$F \rightarrow gFg^{-1}, \tag{39}$$

where $Ad(g^{-1})$ is the adjoint representation of g^{-1}. For a single point as the base space, the bundle space reduces to the Lie group G and the covariant derivative of the connection 1-form vanishes

$$dA + A \wedge A = 0. \tag{40}$$

This equation is called the *Maurer-Cartan structure equation.* The *Maurer-Cartan form* $g^{-1}dg$ belongs to the Lie algebra of the principal bundle. This form is invariant under the left action of a constant group element g_o:

$$(g')^{-1}dg' = (g_o g)^{-1}d(g_o g) = g^{-1}dg. \tag{41}$$

This form can be parametrized as $g^{-1}dg = \phi_a T_a$ where T_a satisfy the algebra (33). Using the identity $d(g^{-1}dg) = -g^{-1}dg \wedge g^{-1}dg$ we obtain

$$d\phi_a + \frac{1}{2} f_{abc} \phi_a \wedge \phi_b = 0. \tag{42}$$

As we move along a curve $x^\mu(\lambda)$ in the base manifold, a corresponding section $S(\lambda)$ (called a *lift*) is traced in the principal bundle according to

$$\frac{d}{d\lambda} = \dot{x}^\mu \frac{\partial}{\partial x^\mu} + \dot{S} \frac{\partial}{\partial S}. \tag{43}$$

The section S is said to be parallel transported if

$$\dot{S}_{ij} + A_{\mu,ik}\dot{x}^{\mu}S_{kj} = 0. \tag{44}$$

From (43) and (44) we obtain

$$\frac{d}{d\lambda} = \dot{x}^{\mu}(\frac{\partial}{\partial x^{\mu}} - A_{\mu}^{a}T_{ij}^{a}S_{jk}\frac{\partial}{\partial S_{ik}}). \tag{45}$$

The expression inside the parentheses gives the *covariant derivative* denoted by D_{μ}:

$$D_{\mu} = \partial_{\mu} - A_{\mu}^{a}T_{a} \tag{46}$$

where $T_a = T_{ij}^{a}S_{jk}\frac{\partial}{\partial S_{ik}}$. The basic motivation for defining the covariant derivative is to modify ∂_{μ} in such a way that the resulting quantity transforms covariantly under the action of the group element g. The components of the curvature 2-form ($F_{\mu\nu}^{a}$) are related to D_{μ} according to

$$[D_{\mu}, D_{\nu}] = -F_{\mu\nu}^{a}T_{a}, \tag{47}$$

or

$$F_{\mu\nu}^{a} = \partial_{\mu}A_{\nu}^{a} - \partial_{\nu}A_{\mu}^{a} + f_{abc}A_{\mu}^{b}A_{\nu}^{c}. \tag{48}$$

In the language of differential forms,

$$\Omega = d\omega + \omega \wedge \omega = g^{-1}Fg, \tag{49}$$

where $\omega = g^{-1}Ag + g^{-1}dg$ is the connection 1-form. Equations (36) and (49) can be combined to obtain

$$F = dA + A \wedge A = \frac{1}{2}F_{\mu\nu}^{a}T^{a}dx^{\mu} \wedge dx^{\nu}. \tag{50}$$

The inhomogeneous equations governing the gauge field read

$$D^{*}F = {}^{*}J, \tag{51}$$

where J is the current 1-form. Conservation of the current J results from the underlying gauge symmetry (Noether's theorem)

$$D^{*}J = 0. \tag{52}$$

Note that it is the dual of the current 1-form which is covariantly closed. The source-free field equations can be obtained from the following action:

$$\mathcal{A} = \int -\frac{1}{4}F \wedge^{*} F. \tag{53}$$

Note that $F \wedge {}^*F$ is a 4-form proportional to the volume 4-form in the 4-dimensional Minkowski space.

In the case of the abelian $U(1)$ gauge symmetry, the group manifold is a circle parametrized by the angle θ ($g = e^{i\theta}$). The group element g acts on the vector space of complex functions ϕ. The principal bundle is locally isomorphic to the product of an open set in the Minkowski space and S^1. The potential 1-form $A = A_\mu dx^\mu$ has real components A_μ in this case. We also have $F = dA = \frac{1}{2} F_{\alpha\beta} dx^\alpha \wedge dx^\beta$ with components as a covariant antisymmetric tensor

$$F_{\alpha\beta} = \partial_\alpha A_\beta - \partial_\beta A_\alpha. \tag{54}$$

Note that $A \wedge A = 0$, since in this case the potential 1-form is not matrix valued and $A_\mu A_\nu dx^\mu \wedge dx^\nu$ vanishes due to the symmetry of $A_\mu A_\nu$ and antisymmetry of $dx^\mu \wedge dx^\nu$. Also note that $A \wedge F = 0$ and the Bianchi identity (36) translates into the following tensorial relation

$$\partial_\alpha F_{\beta\gamma} + \partial_\beta F_{\gamma\alpha} + \partial_\gamma F_{\alpha\beta} = 0, \tag{55}$$

while (51) becomes

$$\partial_\beta F^{\alpha\beta} = J^\alpha. \tag{56}$$

Equations (55) and (56) constitute the complete Maxwell equations.

The $U(1)$ gauge transformation $\phi \rightarrow e^{i\theta}\phi$ leads to

$$A \rightarrow A + d\theta, \quad or \quad A_\mu \rightarrow A_\mu + \partial_\mu \theta \tag{57}$$

while F remains gauge invariant. A section of the principal bundle corresponds to the selection of a particular gauge $\theta(x^\mu)$.

Maxwell's equations imply the nonexistance of magnetic monopoles. The possibility of having magnetic monopoles and its implications has been extensively studied in the literature (see Goddard and Olive, 1978 and also Horva'thy, 1988 for an introductory exposition of the subject).

In the presence of magnetic charges, the homogeneous Maxwell equation is modified as

$$dF = -{}^*K, \tag{58}$$

where K is the magnetic current 1-form. Expressed in tensor components, this equation reads

$$\frac{1}{\sqrt{|g|}} \partial_\mu (\sqrt{|g|}\ {}^*F^{\mu\nu}) = K^\nu, \tag{59}$$

or

$$\partial_\alpha F_{\beta\gamma} + \partial_\gamma F_{\alpha\beta} + \partial_\beta F_{\gamma\alpha} = -\sqrt{|g|}\epsilon_{\alpha\beta\gamma\delta}K^\delta. \tag{60}$$

A point-like monopole located at the origin of spherical coordinates generates the following field

$$F = \frac{g}{4\pi} sin\theta d\theta \wedge d\phi, \tag{61}$$

where g is the strength (magnetic charge) of the monopole. From $F = dA$ the potential 1-form can be chosen as

$$A = -\frac{g}{4\pi} cos\theta d\phi. \tag{62}$$

The total magnetic flux across a sphere centered at the origin is

$$\Phi_B = \oint_{S^2} F = \frac{g}{4\pi}\int_{\phi=0}^{2\pi}\int_{\theta=0}^{\pi} sin\theta d\theta d\phi = g. \tag{63}$$

Note that the gauge potential (63) is *not* smooth everywhere. By computing the components of A in Cartesian coordinates, it is easily seen that the potential diverges along the z-axis. Expressed in mathematical terms, this means that the corresponding bundle is not trivializable and a global section does not exist.

The total angular momentum of the electromagnetic field produced by a static pair of a point charge q and a magnetic monopole g is given by

$$\vec{J} = \int \vec{x} \times (\epsilon_o \vec{E} \times \vec{B}) d^3x = \frac{qg}{4\pi}\hat{k}, \tag{64}$$

where $\hat{k}$ is a unit vector from q to g. Using the quantum mechanical quantization $J = n\frac{\hbar}{2}$, we are led to an explanation for the quantization of the electric charge q in the presence of a magnetic monopole. The magnetic monopole will be considered again in section 8.

Transition functions on principal bundles play the role of gauge transformations. Two fiber coordinates ϕ and ϕ' in $U \cap U'$ transform to each other by $\phi' = g\phi$ where g is the transition function. Under this transformation,

$$A' = gAg^{-1} + gdg^{-1}, \quad \omega' = \omega, \quad F' = gFg^{-1}, \quad and \quad \Omega' = \Omega. \tag{65}$$

In mathematical terms, A and F are called *pullbacks* of ω and Ω, respectively.

The correspondence between the principal bundles and gauge fields can be summarized as follows

Structure group $\leftrightarrow$ Gauge group
Connection pullback (A) $\leftrightarrow$ Gauge potential
Curvature pullback (F) $\leftrightarrow$ Field strength
Associated vector bundles ψ $\leftrightarrow$ Matter fields

Transition functions $\leftrightarrow$ Gauge transformations
Sections $\leftrightarrow$ Gauge fixing conditions
Maurer-Cartan 1-forms $\leftrightarrow$ Pure gauges

The global topology of gauge fields become relevant in the path integral quantization approach. Path integral formalism works properly only for (++++)-signature spaces. This is why the Euclidean signature is of particular importance in the soliton and instanton methods.

For a detailed discussion of gauge theories and differential geometry, the reader is referred to Eguchi et al. (1980).

5 Yang-Mills Field

$SU(2)$ gauge theory was introduced in 1954 by Yang and Mills. The gauge group of the Yang-Mills field is $SU(2)$. Elements of this group are unitary 2×2 matrices which operate on the two-component isospinors ψ. The covariant derivative D_μ is defined in such a way that $D_\mu \psi$ transform in the same way as ψ. According to equation (46)

$$D_\mu = \partial_\mu - A^a_\mu \tau_a, \tag{66}$$

where τ_a are the Pauli matrices[2]. Note that this is a matrix-valued equation and ∂_μ is implicitly assumed to be $I.\partial_\mu$ where I is the 2×2 unit matrix. The tangent space at the group identity $e = I$ defines the Lie algebra of the group. The algebra of group generators τ_a is

$$[\tau_a, \tau_b] = 2i\epsilon_{abc}\tau_c, \tag{67}$$

where ϵ_{abc} is the totally antisymmetric tensor in three dimensions. Note that we have

$$[D_\mu, D_\nu]\psi = iF_{\mu\nu}\psi, \tag{68}$$

where

$$F_{\mu\nu} = \partial_\mu A_\nu - \partial_\nu A_\mu + i[A_\mu, A_\nu] \tag{69}$$

which is the tensor version of equation (35) $F_{\mu\nu} = iF^a_{\mu\nu}\tau_a$ and $A_\mu = A^a_\mu \tau_a$, we have

$$F^a_{\mu\nu} = \partial_\mu A^a_\nu - \partial_\nu A^a_\mu - \frac{1}{2}\epsilon^a{}_{bc}(A^b_\mu A^c_\nu - A^b_\nu A^c_\mu)$$

$$= \partial_\mu A^a_\nu - \partial_\nu A^a_\mu + \epsilon^a{}_{bc} A^b_\mu A^c_\nu. \tag{70}$$

[2]Note that there is some changes of conventions with respect to those used in the previous section

Under an infinitesimal gauge transformation, each of the components of ψ change by $\delta\psi^A$ ($A = 1, 2$).

$$\delta\psi^A = \psi'^A - \psi^A = \epsilon^a (\tau_a)^A_B \psi^B, \tag{71}$$

where ϵ^a (a=1,2,3) are the infinitesimal parameters of the gauge transformations. It can be easily shown that under these transformations A^a_μ and $F^a_{\mu\nu}$ change by

$$\delta A^a_\mu = \epsilon^a{}_{bc} \epsilon^b A^c_\mu + \partial_\mu \epsilon^a, \tag{72}$$

and

$$\delta F^a_{\mu\nu} = \epsilon^a{}_{bc} \epsilon^b F^c_{\mu\nu}. \tag{73}$$

Note that $F^a_{\mu\nu}$ transforms as the adjoint representation of the group $SU(2)$. A *metric* can be defined on the group manifold of $SU(2)$ according to

$$g_{ab} = g_{ba} = \epsilon^c{}_{ad} \epsilon^d{}_{bc} \tag{74}$$

which is an example of the *Cartan-Killing metric*. Note that the isospin indices a, b, are raised and lowered by g^{ab} and g_{ab} in the same manner that the spacetime indices $\mu, \nu, \ldots$ are raised and lowered by the Minkowski metric $\eta_{\mu\nu}$ or $\eta^{\mu\nu}$.

The free-field Lagrangian density of the Yang-Mills field is

$$\mathcal{L}_{free} = -\frac{1}{4} F^a_{\mu\nu} F^{\mu\nu}_a. \tag{75}$$

6 Self-Duality, Instantons, and Monopoles

Solutions of Yang-Mills equations which have the important property

$$F = \pm\, {}^*F \tag{76}$$

are called *self-dual* (+) or *anti-self-dual* (-). In such a case, the equation of motion $D^*F = 0$ reduces to the Bianchi identity $DF = 0$. (Anti)self-dual fields are therefore solutions of the field equations. Among the most interesting solutions of the Yang-Mills equations are instantons and monopoles. The instanton solution of 'tHooft and Polyakov is given by

$$A_\mu = -\frac{1}{g_o} \frac{\eta_{c\mu\nu} \tau^c x^\nu}{r^2 + r_o^2}, \tag{77}$$

where r_o is a constant which prevents A to become singular at $r = 0$. Also $r^2 = \sum_{\mu=1}^4 (x^\mu)^2$, $\eta_{abc} = \epsilon_{abc}$ for $a, b, c = 1, 2, 3$, $\eta_{a4b} = -\delta_{ab}$, $\eta_{ab4} = \delta_{ab}$, and $\delta_{a44} = 0$. At large distances, the solution (77) approaches the asymptotic form

$$A_\mu \to -\frac{i}{g_o} g^{-1} \partial_\mu g, \tag{78}$$

where $g = (r^2 + \lambda^2)^{-1}(x^4 + i\tau_a x^a)$. This asymptotic form is a pure gauge for which the curvature 2-form vanishes:

$$F = DA = dA + A \wedge A = 0. \tag{79}$$

In other words, the potential 1-form is asymptotically of the Maurer-Cartan type. The corresponding gauge function $g(x^\mu)$ is a mapping from the base space (R^4) into the bundle space or the group space. The group manifold of $SU(2)$ is a 3-sphere. As we shall see later, this has important topological implications.

The right action of a group element $h \in G$ on the principal bundle is given by

$$ph = (x, g)h = (x, gh) = p' \tag{80}$$

Note that $\pi(p') = \pi(p) = x$. The vector field associated with the infinitesimal action of the Lie group on the principal bundle is called the *fundamental vector field*

$$\hat{\eta}(p) = \frac{d}{dt}(pe^{t\eta}), \tag{81}$$

or

$$\hat{\eta}(x, g) = (x, g\eta). \tag{82}$$

Matter fields may couple to the Yang-Mills field. For example, the V-valued field ϕ transforms as

$$\phi' = g^{-1}\phi, \tag{83}$$

where $g \in G$. In other words, ϕ is a mapping from the principal bundle P to the vector space V:

$$\phi : P \to V, \quad \phi(pg) = g^{-1}\phi(p) \tag{84}$$

V can be the adjoint representation of G. The covariant derivative of ϕ is given by

$$D\phi = d\phi + e[A, \phi]. \tag{85}$$

The covariant derivative satisfies the Jacobi relation

$$[D_\mu, [D_\nu, D_\lambda]] + [D_\nu, [D_\lambda, D_\mu]] + [D_\lambda, [D_\mu, D_\nu]] = 0. \tag{86}$$

The minimal coupling of the ϕ-field to the Yang-Mills field is implemented in the following Lagrangian density

$$\mathcal{L} = -\frac{1}{4}F \wedge^* F + \frac{1}{2}|D\phi|^2 - U(\phi), \tag{87}$$

where $U(\phi)$ is the potential term for the ϕ-field. The field equation for ϕ reads

$$^*D^*D\phi = -\frac{\partial U}{\partial \phi}, \quad or \quad D_\mu D^\mu \phi^a = -\frac{\partial U}{\partial \phi_a} \tag{88}$$

In terms of components, equation (51) reads

$$D_\mu F^{\mu\nu} = J^\nu. \tag{89}$$

The ϕ-field provides a current density (source) for the Yang-Mills equation

$$J_\mu^a = e[\phi, D_\mu \tau^a \phi]. \tag{90}$$

Note that the underlying symmetry of the Lagrangian demands that J_μ is covariantly conserved ($D_\mu J^\mu = 0$), and at the same time the ordinary conservation law

$$\partial_\mu J^\mu = \partial_\mu \left[\partial_\nu F^{\nu\mu} + e[A_\nu, F^{\nu\mu}]\right) + e\partial_\mu([A_\nu, F^{\mu\nu}])]$$

$$= \partial_\mu \partial_\nu F^{\mu\nu} + e\partial_\mu([A_\nu, F^{\mu\nu} - F^{\nu\mu}]) = 0 \tag{91}$$

is also satisfied. The choice

$$U = \frac{\lambda}{4}(|\phi|^2 - \phi_o^2)^2, \tag{92}$$

where λ and ϕ_o are positive constants leads to the *spontaneous breaking* of the gauge symmetry (*Higgs mechanism*). This mechanism is responsible for the formation of massive vector bosons ($m_A^2 = e^2\phi_o^2$) .

For a static configuration of the Yang-Mills and Higgs fields, the total energy functional becomes

$$E = \int \left[\frac{1}{4}Tr(F_{ij}F^{ij}) + \frac{1}{2}(D_i\phi, fD^i\phi) + U(\phi)\right] d^3x. \tag{93}$$

In 1974, 'tHooft and Polyakov introduced a static, non-singular solution of the Yang-Mills-Higgs equations with remarkable properties. Consider the following *hedgehog* ansatz:

$$\phi^a = G(r)\frac{x^a}{er^2}, \quad and \quad A_i^a = [F(r) - 1]\epsilon_{aij}\frac{x^j}{er^2}, \tag{94}$$

where $F(r)$ and $G(r)$ are unknown functions to be determined by field equations (88) and (89) and $A_o^a = 0$. These field equations lead to the following coupled nonlinear differential equations

$$r^2 \frac{d^2F}{dr^2} = FG^2 + F(F^2 - 1), \tag{95}$$

$$r^2 \frac{d^2G}{dr^2} = 2F^2G + \lambda e^{-2} G(G^2 - \mu^2 e^2 r^2). \tag{96}$$

These equations can also be obtained by applying the variational principle to the energy functional (93).

In order to be single valued at $r = 0$, A_i^a and ϕ^a should vanish at the origin. This demands

$$F(r) \rightarrow 1, \quad and \quad G(r) \rightarrow 0 \quad as \quad r \rightarrow 0. \tag{97}$$

For a localized, finite-energy solution, the gauge field must reduce to a pure gauge (i.e. Maurer-Cartan form) at large r. The ϕ-field must assume its vacuum (i.e. $|\phi| \rightarrow \phi_o$) far from the origin. Therefore,

$$F(r) \rightarrow 0, \quad and \quad G(r) \rightarrow \mu e r \quad as \quad r \rightarrow \infty. \tag{98}$$

This behavior guarantees the vanishing of energy density in (93) as $r \rightarrow \infty$, since in this limit, $F_{ij} \rightarrow 0$, $D\phi \rightarrow 0$, and $U \rightarrow 0$. In other words, although each component of ϕ depends on x^i, the ϕ-field is covariantly "constant".

Note that the vacuum of the ϕ-field corresponds to $U(\phi) = 0$. This defines a 2-sphere in the ϕ_a-space

$$\phi_a \phi_a = (\phi_1)^2 + (\phi_2)^2 + (\phi_3)^3 = \phi_o^2, \tag{99}$$

which is an example of the *manifold of degenerate vacuua*. The asymptotic form of F_{ij}^a at large r corresponds to a radial, inverse square magnetic field

$$B^i = Tr(\epsilon^{ijk} F_{jk} \phi) = -\frac{x^i}{3r^3}. \tag{100}$$

The full solution to equations (95) and (96) can only be obtained numerically. Note that the $SU(2)$ gauge symmetry is spontaneously broken to $U(1)$ at large distances from the monopole (so is the corresponding bundle). The Higgs mechanism leads to the formation of two massive vector fields from the original A_μ^a's. The remaining massless gauge field is interpreted as the ordinary electromagnetic field which leads to the identification of the magnetic field (100). According to 't Hooft, the following gauge invariant quantity properly describes the electromagnetic field

$$F_{\mu\nu} = \hat{\phi}_a F_{a\mu\nu} = -\frac{1}{g} \epsilon_{abc} \hat{\phi}_a D_\mu \hat{\phi}_b D_\nu \hat{\phi}_c, \tag{101}$$

where $\hat{\phi}_a = \phi_a / |\phi|$. With this identification we have

$$ {}^*F^{\mu\nu} = \frac{1}{2g}\epsilon^{\mu\nu\alpha\beta}\hat{\phi}_a\partial_\alpha\hat{\phi}_b\partial_\beta\hat{\phi}_c, \tag{102} $$

and

$$ \partial_\nu\ {}^*F^{\mu\nu} = \frac{4\pi}{g}K^\mu, \tag{103} $$

where K^μ is the magnetic current. The total magnetic charge is given by

$$ Q_m = \frac{1}{g}\int K^0 d^3x = \frac{n}{g}, \tag{104} $$

where n is the topological charge or winding number.

There are exact solutions to the field equations in the $\lambda \to 0$ limit which is known as the *Bogomol'nyi-Prasad-Sommerfield* (BPS) limit (Prasad and Sommerfield, 1975, and Bogomol'nyi, 1976). It can be shown that a self-dual or anti-self-dual field satisfies the Yang-Mills equations. The (anti) self-duality condition ($F = \pm\ {}^*F$) leads to the following first-order equations

$$ F = \pm^* D\phi \quad or \quad F_{ij} = \pm\epsilon_{ijk}D_k\phi. \tag{105} $$

These equations are known as the *Bogomol'nyi equations* . For the hedgehog ansatz (94), equations (105) become the following first order differential equations for $F(r)$ and $G(r)$

$$ r\frac{dG}{dr} = G - (F^2 - 1), \tag{106} $$

$$ r\frac{dF}{dr} = -FG. \tag{107} $$

These equations can be solved exactly. The monopole solution satisfying the appropriate boundary conditions reads

$$ G(r) = \mu er\coth(\mu er) - 1 \quad and \quad F(r) = \frac{\mu er}{\sinh(\mu er)}. \tag{108} $$

The reader can easily verify that these solutions satisfy the equations (106) and (107) by using the change of variables $1+G = \mu er\psi$ and $F = \mu er\chi$. The Bogomol'nyi equations minimize the energy functional (93), with the following total energy

$$ E_{min} = \frac{4\pi\mu|g|}{e}. \tag{109} $$

The (anti)self-dual field configurations are particularly important, since they provide the stationary configurations around which quantum fluctuations can be computed. An important property of these models is that the actions are minimized at values which are proportional to the corresponding topological charges.

If instead of being zero, $A_{a0} = \frac{1}{g}J(r)\frac{x^a}{r^2}$ is assumed, we arrive at the following differential equations

$$r^2\frac{d^2K}{dr^2} = K(K^2 - J^2 + H^2), \tag{110}$$

$$r^2\frac{d^2H}{dr^2} = 2HK^2 + \lambda H\left[\frac{1}{g^2}H^2 - r^2F^2\right], \tag{111}$$

and

$$r^2\frac{d^2J}{dr^2} = 2JK^2, \tag{112}$$

in which the same ansatz for A_{ai} and ϕ_a have been used as before. Solutions to these equations are called *dyons* and they bear both electrical and magnetic charges. The electric charge of a dyon is given by

$$Q_e = -\frac{8\pi}{g}\int\frac{JK^2}{r}dr \tag{113}$$

and is not necessarily quantized. It can be shown that in the BPS limit, $Q_e = \frac{4\pi}{g}\sinh\gamma$ where γ is an arbitrary constant.

7 Topological Currents

Perhaps the simplest example of a topological current is the one associated with the sine-Gordon system. Consider the sine-Gordon Lagrangian density in 1+1 dimensions

$$\mathcal{L}_{SG} = \frac{1}{2}\partial^\mu\phi\partial_\mu\phi - (1 - cos\phi), \tag{114}$$

where ϕ is a real scalar field on a two dimensional spacetime $x^o = t$ and $x^1 = x$ with the metric $\eta_{\mu\nu} = diag(-1, 1)$. The self-interaction potential $V(\phi) = 1 - \cos\phi$ has an infinite number of degenerate vacuua at $\phi_n = 2n\pi$, $n \in Z$. The Lagrangian density (114) leads to the famous sine-Gordon equation

$$\partial^\mu\partial_\mu\phi = sin\phi, \tag{115}$$

which is known to be an integrable equation with a hierarchy of multi-soliton solutions (Lamb, 1980, and Riazi and Gharaati, 1998). Localized, finite-energy solutions of (115) satisfy the following boundary conditions

$$\phi(+\infty) = 2n\pi, \quad \phi(-\infty) = 2m\pi, \tag{116}$$

where m and n are integers. Topological current for the sine-Gordon system is given by

$$J^{\mu} = \frac{1}{2\pi}\epsilon^{\mu\nu}\partial_{\nu}\phi, \tag{117}$$

where $\epsilon^{\mu\nu}$ is the totally antisymmetric tensor in two dimensions. The current density (117) is conserved

$$\partial_{\mu}J^{\mu} = \frac{1}{2\pi}\epsilon^{\mu\nu}\partial_{\mu}\partial_{\nu}\phi = 0, \tag{118}$$

since $\epsilon^{\mu\nu}$ is antisymmetric while $\partial_{\mu}\partial_{\nu}$ is symmetric. The total charge of a localized solution (both static and dynamic) is easily shown to be quantized

$$Q_{SG} = \int_{-\infty}^{+\infty} J^{o}dx = \frac{1}{2\pi}\int_{-\infty}^{+\infty} \epsilon^{o1}\partial_{1}\phi dx$$

$$= \frac{1}{2\pi}\int_{-\infty}^{+\infty}\frac{\partial\phi}{\partial x}dx = \frac{1}{2\pi}\left[\phi(+\infty) - \phi(-\infty)\right] = n - m, \tag{119}$$

in which the boundary conditions (116) are used. Note that solutions with different topological charges belong to distinct topological sectors. They are separated from each other by infinite energy barriers. In other words, they cannot be continuously deformed into each other.

The concept of topological charges can be extended to more complicated fields in higher dimensions. Consider, for example, a time-dependent complex scalar field $\phi = \phi_1 + i\phi_2$ on the complex plane z. The vacuum manifold of ϕ is assumed to reside at

$$|\phi|^2 = \phi_1^2 + \phi_2^2 = \phi_o^2, \tag{120}$$

where ϕ_o is a real, positive constant. The topological current for this field can be defined as

$$J^{\mu} = \frac{1}{2\pi\phi_o^2}\epsilon^{\mu\nu\alpha}\epsilon_{ab}\partial_{\nu}\phi_a\partial_{\alpha}\phi_b, \tag{121}$$

where $\mu, \nu, \alpha = 0, 1, 2$ and $a, b = 1, 2$. It can be easily shown that this current is conserved ($\partial_{\mu}J^{\mu} = 0$). The total charge of a localized field configuration on the complex plane ($z = x + iy$) is

$$Q = \int\int J^{o}dxdy = \frac{1}{2\pi\phi_o^2}\int\int \epsilon^{ij}\epsilon_{ab}\partial_i\phi_a\partial_j\phi_b dxdy. \tag{122}$$

This can be rewritten as $\int\int \vec{\nabla} \times \vec{H}.d\vec{S}$ where $d\vec{S} = dxdy\hat{k}$, and $\vec{\nabla} \times \vec{H}.\hat{k} = \epsilon_{3ij}\partial_i H_j$ with

$$H_j = \frac{\epsilon_{ab}\phi_a\partial_j\phi_b}{2\pi\phi_o^2}. \tag{123}$$

Note that a z-axis is artificially introduced to facilitate the notations of ordinary vector analysis. $\hat{k}$ is the unit vector in the positive z-direction. Stokes's theorem can now be used

$$Q = \oint_C \vec{H}.\vec{dl}, \tag{124}$$

where $\vec{dl}$ is a line element along the boundary curve C which is assumed to be in the form of a circle on the complex plane with radius $r \to \infty$. Parametrizing dl and $d\phi$ along C according to

$$dl = r d\theta, \qquad d\phi = \phi_o d\alpha, \tag{125}$$

leads to

$$Q = \frac{1}{2\pi\phi_o^2} \int_{\theta=0}^{2\pi} \frac{1}{2}(\phi_1 \frac{\partial \phi_2}{\partial \theta} - \phi_2 \frac{\partial \phi_1}{\partial \theta}) r d\theta$$

$$= \int_{\theta=0}^{2\pi} \frac{d\alpha}{d\theta} d\theta = \frac{1}{2\pi} \oint_{\theta=0}^{2\pi} d\alpha(\theta) = n, \tag{126}$$

where n is an integer corresponding to the number of times the ϕ field circles its vacuum S^1 as the path C is completed *once* on the complex plane. We saw in section 3 that n was properly called the winding number.

For an isovector field ϕ_a ($a = 1, 2, 3$) with an S^2 vacuum

$$\phi_a \phi_a = \phi_o^2 \tag{127}$$

the topological current can be defined as (Vasheghani and Riazi, 1996)

$$J^\mu = \frac{1}{4\pi\phi_o^3} \epsilon^{\mu\nu\alpha\beta} \epsilon_{abc} \partial_\nu \phi_a \partial_\alpha \phi_b \partial_\beta \phi_c. \tag{128}$$

Note that the spacetime is now the ordinary Minkowski spacetime and $\mu, \nu, ... = 0, 1, 2, 3$ with $x^o = t$. Once again, the current is identically conserved ($\partial_\mu J^\mu = 0$), and the total charge is quantized

$$Q = \int J^o d^3x = \frac{1}{4\pi\phi_o^3} \oint \frac{dS_\phi}{dS_x} dS_x = n, \tag{129}$$

where dS_ϕ and dS_x are area elements in the x-space and ϕ-space, respectively. The current (128) can be written as the covariant divergence of an anti-symmetric second-rank tensor

$$\partial_\mu F^{\mu\nu} = J^\nu, \tag{130}$$

where

$$F^{\mu\nu} = \frac{1}{4\pi\phi_o^3}\epsilon^{\mu\nu\alpha\beta}\left[\epsilon_{abc}\phi_a\partial_\alpha\phi_b\partial_\beta\phi_c + \partial_\beta\mathcal{B}_\alpha\right], \tag{131}$$

in which $\mathcal{B}_\alpha$ is an auxiliary vector field. It is interesting to note that the dual field $^*F$ with the following tensorial components

$$^*F^{\mu\nu} = \frac{1}{2}\epsilon^{\mu\nu\alpha\beta}F_{\alpha\beta} = 2\epsilon_{abc}\phi_a\partial^\mu\phi_b\partial^\nu\phi_c + \partial^\mu\mathcal{B}^\nu - \partial^\nu\mathcal{B}^\mu \tag{132}$$

satisfies the equation

$$\partial_\mu\ {}^*F^{\mu\nu} = 0, \tag{133}$$

provided that the vector field $\mathcal{B}^\mu$ is a solution of the following wave equation

$$\Box\mathcal{B}^\mu - \partial^\mu(\partial_\alpha\mathcal{B}^\alpha) = 2\epsilon_{abc}\partial_\alpha(\phi_a\partial^\mu\phi_b\partial^\alpha\phi_c). \tag{134}$$

The right hand side of this equation defines another conserved current

$$K^\mu = 2\epsilon_{abc}\partial_\alpha(\phi_a\partial^\mu\phi_b\partial^\alpha\phi_c), \tag{135}$$

which is consistent with the vanishing of the divergence of the left hand side of equation (134). The resemblance of equations (130) and (133) to Maxwell's equations and the capability of this model to provide non-singular models of charged particles is discussed in Vasheghani and Riazi (1996). Let us write (131) in the following form

$$G = F - H, \tag{136}$$

where

$$G^{\mu\nu} = \frac{1}{4\pi\phi_o^3}\epsilon^{\alpha\beta\mu\nu}\epsilon_{abc}\phi_a\partial_\alpha\phi_b\partial_\beta\phi_c \tag{137}$$

and

$$H^{\mu\nu} = \frac{1}{4\pi\phi_o^3}\epsilon^{\mu\nu\alpha\beta}\partial_\beta\mathcal{B}_\alpha. \tag{138}$$

We now have

$$dF = 0,$$

and

$$d^*H = 0. \tag{139}$$

Any 2-form like G which can be written as the sum of two parts satisfying (139) is said to admit *Hodge decomposition* . We shall see in the next section that forms like G and H are cohomologous (i.e. they belong to the same cohomology class), since they differ only by an exact form.

Topological currents of defects in various space dimensions can be formulated in a unified way. Duan et al. (1999) considered the following topological current for point defects in a medium represented by an $n-$dimensional order parameter ϕ^a ($a = 1, ..., n$):

$$J^{\mu} = \frac{1}{A(S^{n-1})(n-1)!}\epsilon^{\mu\mu_1...\mu_n}\epsilon_{\hat{\phi}_1...\hat{\phi}_n}\partial_{\mu_1}\hat{\phi}^{a_1}...\partial_{\mu_n}\hat{\phi}^{a_n}, \tag{140}$$

where $A(S^{n-1}) = \frac{2\pi^{n/2}}{\Gamma(n/2)}$ is the area of the $(n-1)$-dimensional unit sphere, and

$$\hat{\phi}^a = \frac{\phi^a}{|\phi|}. \tag{141}$$

The number of space dimensions is also assumed to be n. The topological charge density $\rho = J^o$ which represents the defect density, is everywhere zero except at the location of the defects where it diverges. It therefore behaves like a delta function, and can be represented as

$$J^{\mu} = \delta(\phi)D^{\mu}(\frac{\phi}{x}), \tag{142}$$

where $\delta(\phi)$ is the Dirac delta function, and the $n-$dimensional Jacobian determinant is defined by

$$\epsilon^{a_1...a_n}D^{\mu}(\frac{\phi}{x}) = \epsilon^{\mu\mu_1...\mu_n}\partial_{\mu_1}\phi^{a_1}...\partial_{\mu_n}\phi^{a_n}. \tag{143}$$

The defects are therefore located at points where the equations

$$\phi^a = 0 \tag{144}$$

are satisfied. The defect density is given by (Liu and Mazenko, 1997):

$$\rho = J^o = \delta(\phi)D^o(\frac{\phi}{x}), \tag{145}$$

where $D^o = |\frac{\partial(\phi^1...\phi^n)}{\partial(x^1,...,x^n)}|$ is the ordinary Jacobian determinant. The total topological charge is given by

$$Q = \int J^o d^n x = \sum_{i=1}^{l}\beta_i\eta_i, \tag{146}$$

where β_i and η_i are the Hopf indices and Brouwer degrees of the ϕ mapping, respectively, and l is the number of point defects in the system.

8 Cohomology and Electromagnetism

The theory of cohomology groups was developed by G. de Rham in 1930's. In this section we present a brief introduction to the subject. The relevance of cohomological methods to other gauge fields is discussed in (Henneaux, 1988).

Closed $p-$forms on a manifold $\mathcal{M}$ form a vector space denoted by $C^p(\mathcal{M})$. These forms are also called *p-cocycles.* The subspace of $C^p(\mathcal{M})$ which comprises exact $p-$forms is denoted by $B^p(\mathcal{M})$. Exact and closed $p-$forms are called *p-coboundaries* . The p-th cohomology group is defined as the quotient group $C^p(\mathcal{M})/B^p(\mathcal{M})$:

$$H^p(\mathcal{M}) = C^p(\mathcal{M})/B^p(\mathcal{M}). \tag{147}$$

The elements of $H^p(\mathcal{M})$ form equivalence classes which are called *p-th cohomology classes.*

Two closed $p-$forms ω and η are *cohomologous* (i.e. belong to the same cohomology class) if their difference is an exact form

$$\omega - \eta = d\rho, \tag{148}$$

where ρ is a $(p-1)-$form. The p-th de Rham cohomology group is the set of these equivalence classes. Note that the 0-cohomology class is the class of exact $p-$forms (in other words, all exact $p-$forms are cohomologous).

Let us list a few well-known results relevant to the cohomology groups (Guillemin and Pollack, 1974)

- Exact 0-forms do not exist.
- $H^p(R \times M)$ is isomorphic to $H^p(M)$.
- $H^p(R^k) = 0$ if $k > 0$ and $p > 0$. Every closed $p-$form ($p > 0$) on R^k ($k > 0$) is exact.
- The previous item can be generalized to $H^p(\mathcal{M}) = 0$ for all $p > 0$ if $\mathcal{M}$ is contractible.
- $H^p(S^k)$ is one dimensional for $p = 0$ and $p = k$. For all other $k > 0$, $H^p(S^k) = 0$.
- $H^o(\mathcal{M})$ is the space of constant functions on $\mathcal{M}$ and its dimension counts the number of connected pieces of the manifold. (e.g. $dim H^o(R^n) = 1$). Manifolds which have globally trivial coordinates (*e.g.* R^n) have trivial de Rham cohomologies ($H^p(R^n) = 0$ for $p > 0$).
- The *Euler-Poincare' characteristic* of the manifold $\mathcal{M}$ is defined by

$$\chi_{\mathcal{M}} = \sum_{p=0}^{n} (-1)^p d_p, \tag{149}$$

where $d_p = dim H^p(\mathcal{M})$ is the dimension of the $p-$th homology group (called the *Betti number*). For an $n-$dimensional sphere S^n, $H^o(S^n) = H^n(S^n) = R$ and $H^p(S^n) = 0$ for $0 < p < n$. We therefore have $\chi(S^n) = 0$ for $n = odd$ and $\chi(S^n) = 2$ for $n = even$. For odd-dimensional manifolds, the Euler characteristic vanishes as we saw in the above example.

In the absence of magnetic monopoles $dF = 0$ where F is the electromagnetic field 2-form. F is therefore a closed 2-form. It is also exact and can be written as $F = dA$ where A is the electromagnetic potential 1-form. The phase factor $exp(i \oint_c A)$ which appears in the quantum mechanical context (e.g. the Aharonov-Bohm effect) can be written as

$$exp(i \oint_C A) = exp(i \int_S F), \tag{150}$$

using the Stokes's theorem. Here, S is a surface area with the closed boundary curve C. But $\int_S F$ is the magnetic flux passing through S . The phase factor (150) is known to be observable as shifts in in the interference fringes of electrons in the Aharonov-Bohm experiment. This shows that although A is ambiguous up to a $U(1)$ gauge transformation, it cannot be assumed redundant at a quantum level. In the presence of magnetic monopoles, F is no longer closed. A single-valued gauge transformation in the region surrounding a magnetic monopole (over which F is closed), leads to a Dirac relation between the electric and magnetic charges (Wu and Yang, 1975)

$$g = \frac{1}{e}. \tag{151}$$

In section 4 we saw that the field of a magnetic monopole can be obtained from the potential 1-form

$$A = -\frac{g}{4\pi} cos\theta d\phi. \tag{152}$$

Although this potential looks smooth over a 2-sphere, the coordinate system itself is singular along the z-axis. This gauge potential, therefore, is not defined on the z-axis. When transformed into Cartesian coordinates, this potential becomes

$$-\frac{g}{4\pi}\left(-\frac{zy}{r(x^2+y^2)}, +\frac{zx}{r(x^2+y^2)}, 0\right),$$

which clearly shows the singularity along the z-axis.
The gauge potential of a magnetic monopole can be chosen as

$$A_\pm = \frac{n}{2r}\frac{xdy - ydx}{z \pm r}, \tag{153}$$

where $A_\pm$ are the potentials over the northern (+) and southern (-) hemispheres of an S^2 centered on the monopole. Over an equatorial strip, $A_\pm$ are related by $A_+ = A_- + nd\phi$ which shows that A_+ and A_- do not merge smoothly unless $n = 0$. Note that $\oint_{S^2} F = 2n\pi$ and therefore n represents the quantized magnetic charge of the monopole. The monopole singularity at $r = 0$ should be considered as a hole in the base manifold. The magnetic charge arises from the non-trivial topology of the principal bundle. Occurrence of such integers associated with non-trivial bundles are properly described by the concept of characteristic classes, to be discussed briefly in the section 10.

Electric and magnetic charges act as holes in the base manifold (Minkowski spacetime in this case). Outside these holes, F and $^*F$ are closed and their integral over a closed surface are integer multiples of elementary electric and magnetic charges. In fact, even in the absence of magnetic and electric charges, topologically non-trivial curved spaces can lead to similar effects. Consider, for example, the source-free Maxwell's equations on the following so-called *wormhole* spacetime

$$d\tau^2 = dt^2 - dr^2 - (r^2 + r_o^2)(d\theta^2 + \sin^2\theta d\phi^2), \tag{154}$$

in which $r_o > 0$ is called the throat radius (Morris and Thorne, 1988). Note that r extends from $-\infty$ to $+\infty$, and the metric (154) describes two asymptotically Minkowskian spacetimes joined by an S^2. Covariant Maxwell's equations $F^{\mu\nu}_{;\mu} = 0$ and $^*F^{\mu\nu}_{;\mu} = 0$ have the following non-singular solutions

$$E(r) = \frac{Q_E}{r^2 + r_o^2}; \qquad B(r) = \frac{Q_M}{r^2 + r_o^2}, \tag{155}$$

where E and B are the radial electric and magnetic fields and Q_E and Q_M are constants of integration. The Stokes's theorem $\int_V dF = \int_{\partial V} F$ and a similar relation for $^*F$, when applied to either side of the wormhole imply that the electric and magnetic fluxes through any 2-sphere centered at the wormhole are independent of r. Here, V is a 3-volume confined between two parallel 2-spheres.

9 Homotopy Groups and Cosmic Strings

Let C_1 and C_2 be two loops in a manifold $\mathcal{M}$, based at a point x_o. These two loops are *homotopic* (denoted by $C_1 \simeq C_2$) if they can be continuously deformed into each other. This corresponds to the existence of a continuous set of curves $C(\lambda)$ with $\lambda \in [0, 1]$ such that $C(0) = C_1$ and $C(1) = C_2$. Homotopy is an equivalence relation and the set of all closed curves in $\mathcal{M}$ are divided into a set of homotopy classes. The composite curve $C_3 = C_2 \circ C_1$ is defined according to

$$C_3(\lambda) = \begin{cases} C_2(2\lambda) & if \quad 0 \leq \lambda \leq 1/2, \\ C_1(2\lambda - 1), & if \quad 1/2 \leq \lambda \leq 1. \end{cases} \tag{156}$$

Homotopic deformations of any class C provide the identity (Id) of that class. The inverse of a homotopy C^{-1} is defined according to the relation $C^{-1} \circ C = C \circ C^{-1} = Id$. A loop is homotopic to zero if it can be continuously deformed into a point. Such loops are also called *null-homotopic.* Homotopy classes with the inverse and identity defined in this way form a group called the first or *fundamental group* denoted by π_1. A manifold is called *simply connected* if all loops in it are null-homotopic. R^n is an obvious example of a simply connected manifold ($\pi_1(R^n) = 0$). The group manifold of $U(1)$ is a circle (S^1). The fundamental group of $U(1)$ therefore corresponds to the number of times a loop circles the group manifold in the clockwise or counter-clockwise direction. We therefore have $\pi_1(S^1) = Z$, and the same is for $R^2 - \{0\}$. Other examples of fundamental groups include $\pi_1(O(2)) = \pi_1(SO(2)) = \pi_1(U(N)) = Z$, while $\pi_1(SO(3)) = \pi_1(SO(N)) = \pi_1(O(N)) = Z_2$ ($N \geq 3$), where Z_2 is the group of integers modulo 2. Furthermore, $SU(N)$ groups are simply connected and their fundamental group vanishes.

Higher homotopy groups are defined in a similar manner using compact hypersurfaces homotopic to S^n (i.e. continuously deformable to an $n-$dimensional sphere). The $n-$th homotopy group π_n therefore comprises the homotopy classes of maps from S^n to the manifold under consideration. In particular, we have $\pi_n(S^n) = Z$, $\pi_3(S^2) = Z$, and since the group manifold of $SU(2)$ is S^3, $\pi_3(SU(2)) = Z$.

A manifold $\mathcal{M}$ is *p-connected* if all homotopy groups $\pi_i(\mathcal{M})$ vanish for $i \leq p$.

The wormhole space considered in the previous section is simply connected but not 2-connected, since 2-spheres which contain the wormhole cannot be contracted to a point. Homotopy groups have important implications for the existence and stability of topological defects and solitons. Consider, for example, a complex scalar field ϕ with the self-interaction potential

$$V(\phi) = \frac{\lambda}{4}(\phi^*\phi - \phi_o^2)^2, \tag{157}$$

where λ and ϕ_o are constants. The vacuum manifold is a S^1 on the complex ϕ plane. We can now re-interpret the contents of section 7 (equation 121 onwards) in the framework of homotopy groups. For a localized field on the xy-plane, the $\phi(x, y)$ field along a large circle $x^2 + y^2 = r^2$ ($r \to \infty$) belongs to a homotopy class mapping the circle $S^1 : x^2 + y^2 = r^2$ into the circle $S^1 : \phi_1^2 + \phi_2^2 = \phi_o^2$. This is nothing but the fundamental group S^1: $\pi_1(S^1) = Z$. The topological charge (126) corresponds to the degree of this mapping and labels the corresponding homotopy class. However, the Lagrangian density

$$\mathcal{L} = \partial^\mu \phi^* \partial_\mu \phi - \frac{\lambda}{4}(\phi^*\phi - \phi_o^2)^2 \tag{158}$$

does not lead to localized finite energy solutions. This problem can be demonstrated by the $Q = 1$ sector with the asymptotic behavior $\phi = \phi_o \exp(i\theta)$ in which $\theta = \arctan(y/x)$. The presence of the $|\nabla\phi|^2$ term in the Hamiltonian density

$$\mathcal{H} = \frac{1}{2}|\nabla\phi|^2 + V(\phi), \tag{159}$$

leads to an energy density

$$u_\phi \simeq |\nabla\phi|^2 = \phi_o^2(\nabla\theta)^2 = \frac{\phi_o^2}{r^2}, \tag{160}$$

in plane polar coordinates. This, however, leads to a logarithmically divergent energy integral

$$\int^\infty \frac{\phi_o^2}{r^2} 2\pi r dr = \phi_o^2 \ln r|^\infty \to \infty. \tag{161}$$

In order to cure this problem, the global $U(1)$ symmetry of the Lagrangian (159) can be made local

$$\mathcal{L} = D^\mu\phi^* D_\mu\phi - \frac{\lambda}{4}(\phi^*\phi - \phi_o^2)^2 - \frac{1}{4}F^{\mu\nu}F_{\mu\nu}. \tag{162}$$

This leads to the so-called the *Abelian Higgs model.* We demand that the ϕ field be covariantly constant at large r

$$D\phi = 0 \quad \to \vec{\nabla}\phi - ie\vec{A} = 0, \tag{163}$$

with the asymptotic value $\phi = \phi_o \exp(i\theta)$. Equation (163) gives

$$\vec{A} = \vec{\nabla}(-\frac{i\phi}{e}) \quad as \quad r \to \infty. \tag{164}$$

The potential 1-form A is therefore a pure gauge and leads to a vanishing curvature F at large distances. The energy density of the gauge field, therefore, also vanishes at large distances. Such a solution is non-trivial, since it corresponds to a non-vanishing magnetic flux through the xy-plane

$$\Phi_B = \int F = \int dA = \oint A = \oint \vec{\nabla}(-\frac{i\phi}{e}).\vec{dl} = \frac{1}{e}\phi_o \oint d\theta = \frac{2\pi}{e}\phi_o, \tag{165}$$

where the integration is performed over an infinitely large disk. Solutions belonging to other homotopy classes with topological charges (126) have magnetic fluxes $\Phi_B = \frac{2\pi n}{e}\phi_o$. These solutions which are in the form of bundles of magnetic lines of force in a three dimensional space are called *cosmic strings* in cosmological terminology. They can also represent magnetic flux tubes in the Landau-Ginzburg model of superconductivity.

For a cosmic string, we have a smooth map from the circle S^1 in the configuration space to the $U(1)$ manifold which was also S^1. The Brouwer degree or winding number of this map corresponds exactly to the integers which label the first homotopy group $\pi_1(S^1)$. In this case, the winding number is given by

$$n = \frac{1}{2\pi\phi_o^2} <^* d\phi|id\phi> . \tag{166}$$

Note that (Felsager, 1983)

$$<^* d\phi|id\phi> = -i\int_{R^2} d\bar{\phi}\wedge d\phi = -i\int_{R^2} d(\bar{\phi}d\phi)$$

$$= \lim_{r_o\to\infty} -i\int_{\rho=\rho_o} \bar{\phi}d\phi = -i\phi_o^2 \lim_{\rho_o\to\infty}\int_{\rho=\rho_o} id\phi = 2n\pi\phi_o^2, \tag{167}$$

in which $\bar{\phi}$ is the complex conjugate of ϕ.

The model described above can also be applied to the magnetic flux tubes which form in superconductors. Electrons in a superconductor form pairs. These so-called *Cooper pairs* can be described by a complex scalar field ϕ which is called the *order parameter.* $|\phi|^2$ represents the density of the Cooper pairs. The energy density of a static configuration is given by

$$\mathcal{H} = \frac{1}{2}|\vec{\nabla}\phi|^2 + \frac{1}{2}\alpha|\phi|^2 + \frac{1}{4}\beta|\phi|^4 + C, \tag{168}$$

where C is a constant and α is a parameter which depends on the temperature via

$$\alpha(T) = a\frac{T - T_c}{T_c}, \tag{169}$$

in which a is a positive constant and T_c is the critical temperature. For $T > T_c$, the state of minimum energy density happens at $\phi = 0$ while for $T < T_c$, the minimum converts into a S^1 in the $(Re(\phi), Im(\phi))$ plane

$$|\phi|^2 = -\frac{\alpha}{\beta} > 0. \tag{170}$$

The quantity $\xi(T) \equiv \frac{1}{\sqrt{-\alpha}}$ has the dimensions of length and is called the *coherence length* of the superconductor. In the presence of a magnetic field, we have to implement the covariant derivative

$$D = d - i\frac{g}{\hbar}A, \tag{171}$$

where A is the magnetic vector potential, and include the EM energy density in (168).

It can be shown that $\vec{B}$ satisfies

$$\nabla^2 \vec{B} + \frac{\alpha q^2}{\beta \hbar^2} \vec{B} = 0, \tag{172}$$

showing that the magnetic field is exponentially damped inside a superconductor. The quantity $\sqrt{-\alpha/\beta}q/\hbar$ is called the *penetration length* of the superconductor. The reader notes that the Abelian Higgs model is mathematically equivalent to the Ginzburg-Landau theory of superconductivity. The following correspondence can be stablished between the parameters of the two models

$$e \leftrightarrow g/\hbar, \quad \lambda \leftrightarrow \beta, \quad and \quad \mu^2 \leftrightarrow -\alpha. \tag{173}$$

It can be visualized that a pair of magnetic monopoles with opposite charges may reside at the end-points of a finite string. The quantization of magnetic charge is consistent with the quantization of the magnetic flux through the string ($\Phi_B = g = \frac{2\pi n}{e}$). Since any increase in the distance between the monopole pair is associated with an equal increase in the length of the string, a linearly increasing potential between the monopoles is implied (Felsager, 1983)

$$V(r) = \Lambda r, \tag{174}$$

where Λ is the mass per unit length of the string. This model provides a possible mechanism for the confinement of magnetic monopoles. A similar mechanism was suggested for the quark confinement inside the hadrons (Nambu, 1985).

A rotating relativistic string has the interesting property that its angular momentum J is proportional to its mass squared (M^2). This property is observed in the *Regge trajectories* of the baryons with the same isospin and strangeness. A simpler realization of confinement for solitons with fractional topological charges was introduced by Riazi and Gharaati (1998).

10 Characteristic Classes

In dealing with non-trivial fiber bundles, transition functions and integrals of the curvature 2-form lead to integers which have a topological origin. Characteristic classes are an efficient and elegant way to distinguish and classify inequivalent fiber bundles.

Consider a $k \times k$ complex matrix m. A polynomial $P(m)$ constructed with the components of m is a *characteristic polynomial* if

$$P(m) = P(g^{-1}mg) \quad for \quad all \quad g \in GL(k, C), \tag{175}$$

where g represents a complex matrix belonging to general linear transformations. For example,

$$\det(1 + m) = 1 + S_1(\lambda) + ... + S_k(\lambda) \tag{176}$$

is an invariant polynomial constructed with the i–th symmetrical polynomial

$$S_j(\lambda) = \sum_{i_1<...<i_j} \lambda_{i_1}...\lambda_{i_j}, \qquad (177)$$

where λ_is are the eigenvalues of m. Curvature 2-forms Ω which are matrix-valued have $P(\Omega)$ which are closed and have invariant integrals (Chern, 1967). The *total Chern form* is defined as

$$C(\Omega) = \det(1 + \frac{i}{2\pi}\Omega) = 1 + c_1(\Omega) + c_2(\Omega) + ... \qquad (178)$$

where $c_i(\Omega)$ is a polynomial of degree i in the curvature 2-form Ω:

$$c_o = 1; \qquad c_1 = \frac{i}{2\pi} Tr\Omega,$$

$$c_2 = \frac{1}{8\pi^2}\{Tr(\Omega \wedge \Omega) - Tr\Omega \wedge Tr\Omega\}, \quad etc. \qquad (179)$$

Note that $c_i = 0$ for $2i > n$ where n is the dimension of the base manifold. Closedness causes the Chern forms $c_i(\Omega)$ belong to distinct cohomology classes. These classes have integer coefficients. Integrals like

$$\int_{\mathcal{M}} c_2(\Omega) \quad and \quad \int_{\mathcal{M}} c_1(\Omega) \wedge c_1(\Omega) \qquad (180)$$

are invariant integers called *Chern numbers* .

The action for (anti)self-dual field configurations is proportional to the second Chern number

$$S = -\frac{1}{2}\int TrF \wedge^* F = \mp\frac{1}{2}\int TrF \wedge F = 4\pi|C_2|, \qquad (181)$$

where $C_2 = \frac{1}{8\pi}\int TrF \wedge F$ is the second Chern number. The $|C_2| = 1$ case corresponds to the 't Hooft instanton.

The Chern form is a global form on the base manifold and does not depend on the frame chosen. Chern classes are also closely related to the homotopy theory, since the set of isomorphic classes of i–dimensional vector bundles is isomorphic to the homotopy classes of maps from the base manifold to $Gr(m, i, C)$ where $Gr(m, i, C)$ is the Grassmann manifold of i-planes in C^m.

As a simple example, consider the $U(1)$ bundle of the Dirac monopole over a S^2. We have

$$det(1 + \frac{i}{2\pi}\Omega) = 1 + \frac{i}{2\pi}\Omega \qquad (182)$$

and therefore $c_1 = i\Omega/2\pi$. Note that c_1 is real since Ω is pure imaginary ($\Omega = iF = idA$). Furthermore, using equation (153) we have

$$\int_{S^2} c_1 = -\frac{1}{2\pi}\int_{S^2_+} dA_+ - \frac{1}{2\pi}\int_{S^2_-} dA_- = -\frac{1}{2\pi}\int_o^{2\pi} nd\phi = -n, \tag{183}$$

which shows that the Chern number and the monopole charge are essentially the same for this example. Note that the integration over S^2 is divided into two hemispheres S^2_+ and S^2_- and $dA_+ - dA_- = nd\phi$ is used.

The characteristic classes of real vector bundles are called *Pontrjagin classes.* In similarity with the definition (178), the *total Pontrjagin class* of a real $O(k)$ bundle is defined as

$$P(\Omega) = \det(1 - \frac{1}{2\pi}\Omega) = 1 + p_1 + p_2 + ..., \tag{184}$$

where Ω is the bundle curvature. The orthogonality conditions lead to the vanishing of odd-degree polynomials. The invariant polynomials are closed here also and the resulting cohomology classes are independent of the connection form.

One can also attribute Pontrjagin classes to the electromagnetic field in the following way (Eguchi et al., 1980)

$$\det(1 - \frac{1}{2\pi}F) = 1 + p_1 + p_2, \tag{185}$$

where F is the electromagnetic field tensor in its matrix form, and p_1 and p_2 are related to the EM energy density and Poynting vector according to $p_1 = \frac{1}{(2\pi)^2}(E^2 + B^2)$ and $p_2 = \frac{1}{(2\pi)^4}(\vec{E}.\vec{B})^2$.

11 Differential Geometry and Riemannian Manifolds

We closely follow Eguchi et al. (1980) in this section. The metric on a Riemannian manifold $\mathcal{M}$ can be written in the form

$$ds^2 = g_{\mu\nu}dx^\mu dx^\nu = e^a \eta_{ab} e^b, \tag{186}$$

where $e^a = e^a_\mu dx^\mu$ is the vierbein basis of $T^*(\mathcal{M})$ and η_{ab} is the flat metric ($\eta_{ab} = \delta_{ab}$ for a Euclidean manifold). The connection 1-form $\omega^a_b = \omega^a_{b\mu}dx^\mu$ obeys the Cartan structure equations

$$T^a = de^a + \omega^a_b \wedge e^b, \tag{187}$$

$$R^a_b = d\omega^a_b + \omega^a_c \wedge \omega^c_b, \tag{188}$$

where $T^a = \frac{1}{2}T^a_{bc}e^b \wedge e^c$ is the torsion 2-form and $R^a_b = \frac{1}{2}R^a_{bcd}e^c \wedge e^d$ is the curvature 2-form. The Cartan identities read

$$dT^a + \omega^a_b \wedge T^b = R^a_b \wedge e^b, \tag{189}$$

and

$$dR^a_b + \omega^a_c \wedge R^c_b - R^a_c \wedge \omega^c_b = 0, \tag{190}$$

which is nothing but the Bianchi identity.
In tensor components,

$$T^\mu_{\alpha\beta} = \frac{1}{2}(\Gamma^\mu_{\alpha\beta} - \Gamma^\mu_{\beta\alpha}), \tag{191}$$

where

$$\Gamma^\mu_{\alpha\beta} = \frac{1}{2}g^{\mu\nu}(g_{\nu\beta,\alpha} + g_{\nu\alpha,\beta} - g_{\alpha\beta,\nu}), \tag{192}$$

are the Christoffel symbols, and

$$R^a_b = \frac{1}{2}R^a_{b\mu\nu}dx^\mu \wedge dx^\nu, \tag{193}$$

where

$$R^\alpha_{\beta\mu\nu} = \partial_\mu\Gamma^\alpha_{\nu\beta} - \partial_\nu\Gamma^\alpha_{\mu\beta} + \Gamma^\alpha_{\mu\gamma}\Gamma^\gamma_{\nu\beta} - \Gamma^\alpha_{\nu\gamma}\Gamma^\gamma_{\mu\beta} \tag{194}$$

is the Riemann tensor. As a simple example, consider the 2-sphere

$$ds^2 = r_o^2(d\theta^2 + \sin^2\theta d\phi^2) = (e^1)^2 + (e^2)^2, \tag{195}$$

where r_o is the radius of the sphere, and $e^1 = r_o d\theta$ and $e^2 = r_o sin\theta d\phi$. The structure equations reduce to

$$de^1 = -\omega^1_2 \wedge e^2 = 0 \quad and \quad de^2 = -\omega^2_1 \wedge e^1 = r_o cos d\theta \wedge d\phi, \tag{196}$$

where $\omega^1_2 = -cos\theta d\phi$ is the connection 1-form. Note that $T^a = 0$ here. The curvature 2-form becomes

$$R^1_2 = R^1_{212}e^1 \wedge e^2. \tag{197}$$

Note also that

$$R^1_2 = d\omega^1_2 = \frac{1}{r^2}e^1 \wedge e^2. \tag{198}$$

The volume form on S^2 is $\Omega = r_o^2 \sin\theta d\theta \wedge d\phi$ with $\int \Omega = 4\pi r_o^2$.

Metrics can also be defined on the group manifolds like those of $SU(N)$. A metric on G is defined by the inner product

$$(g'(0), h'(0)) = -Tr(g_o^{-1}g'(0)g_o^{-1}h'(0)), \tag{199}$$

where $g(t)$ and $h(t)$ are two curves in G which have an intersection at $t = 0$:

$$g(0) = h(0) = g_o. \tag{200}$$

This metric is positive definite and multiplication on right or left correspond to isometries of the metric.

Betti numbers (b_m) can also be ascribed to the Riemannian manifolds. Compact orientable manifolds obey the *Poincare' duality* which states that H^p is dual to H^{n-p}. This implies $b_p = b_{n-p}$ which for a four-dimensional manifold becomes $b_o = b_4$ and $b_1 = b_3$. Note that $b_0 = b_4$ counts the number of disjoint pieces of $\mathcal{M}$, and $b_1 = b_3$ vanishes if the manifold is simply connected. Recall that $\chi = b_o - b_1 + b_2 - b_3 + b_4$ is the Euler characteristic of the manifold. The following is an important theorem which relates the local curvature and the global characteristics of hypersurfaces in R^n:

Gauss-Bonnet Theorem: For any compact, even-dimensional hypersurface S in R^{n+1},

$$\int_S \kappa = \frac{1}{2}\gamma_n \chi(S), \tag{201}$$

where κ is the curvature of S (see below), $\chi(S)$ is the Euler characteristic of S and γ_n is the volume of the unit $n-$sphere given by

$$V_n = \frac{2(\pi)^{\frac{n+1}{2}}}{\Gamma(\frac{n+1}{2})}, \tag{202}$$

where Γ is the gamma function ($\Gamma(1/2) = \sqrt{\pi}$, $\Gamma(1) = 1$, and $\Gamma(x+1) = x\Gamma(x)$).

In a curved manifold, the Hodge * operation involves the metric determinant g and $\epsilon_{\mu\nu\dots} = g\epsilon^{\mu\nu\dots}$:

$$^*(dx^{\mu_1} \wedge \dots \wedge dx^{\mu_p}) = \frac{\sqrt{|g|}}{(n-p)!}\epsilon^{\mu_1\dots\mu_p}_{\mu_{p+1}\dots\mu_n} dx^{\mu_{p+1}} \wedge \dots \wedge dx^{\mu_n}. \tag{203}$$

For a scalar field ϕ on an $n-$dimensional Riemannian manifold, $^*\phi = \sqrt{|g|}\phi dx^1 \wedge \dots \wedge dx^n$. We also have $\epsilon = {}^*1$, $^{**}T = (-1)^{k(n-k)}T$ for Euclidean signature and $^{**}T = -(-1)^{k(n-k)}T$ for Minkowskian signature.

Torus $T^2 = S^1 \times S^1$ is a compact 2-dimensional manifold which can be covered by the coordinates $0 \leq (\theta_1, \theta_2) \leq 2\pi$. A torus can not be covered with coordinates which are globally smooth. For example, the θ_1 and θ_2 coordinates are discontinuous at the identified circles $\theta_i = 0$ and $\theta_i = 2\pi$. The 1-forms $d\theta_1$ and $d\theta_2$ are therefore not exact. These two 1-forms provide a basis for the first cohomology group of T^2, with $\dim H^1(T^2) = 2$. Also $H^2(T^2)$ has the basis $d\theta_1 \wedge d\theta_2$ and $dim H^2(T^2) = 1$. We therefore have

$$\sum_{p=0}^{2}(-1)^p dim H^p(T^2) = +1 - 2 + 1 = 0, \tag{204}$$

$$\sum_{p=0}^{2}(-1)^p dim H^p(T^2) = +1 - 2 + 1 = 0, \tag{204}$$

which is equal to the Euler characteristic of the torus.

Consider an oriented $n-$dimensional hyper-surface X in R^{n+1}. The outward pointing normal $\hat{n}(x)$ to this hyper-surface maps X into an n-sphere S^n and is called the *Gauss map*:

$$g : X \to S^n. \tag{205}$$

The Jacobian of this map $J_g(x)$ is called the curvature of X at x and is denoted by $\kappa(x)$. For an n-sphere, $\kappa = \frac{1}{r^n}$ independent of the point x on S^n.

12 Two Dimensional Ferromagnet

Consider a three-component scalar field ϕ_a (a=1,2,3) with the following $O(3)$ Lagrangian density (Belavin and Polyakov 1975, Rajaraman 1988)

$$\mathcal{L} = \frac{1}{2}\partial^\mu \phi_a \partial_\mu \phi_a. \tag{206}$$

The three components of the scalar field are constrained to the surface of a sphere :

$$\phi_a \phi_a = \phi_1^2 + \phi_2^2 + \phi_3^2 = 1. \tag{207}$$

Using the method of Lagrangian multipliers, the corresponding field equation is found to be

$$\Box \phi_a - (\phi_b \Box \phi_b)\phi_a = 0. \tag{208}$$

Static configurations on the xy-plane are described by the following equation

$$\nabla^2 \phi_a - (\phi_b \nabla^2 \phi_b)\phi_a = 0. \tag{209}$$

Note that

$$\phi_a \phi_a = 1 \quad \to \quad \phi_a \vec{\nabla} \phi_a = 0$$

$$\to \quad \nabla.(\phi_a \vec{\nabla} \phi_a) = \vec{\nabla}\phi_a.\vec{\nabla}\phi_a + \phi_a \nabla^2 \phi_a = 0$$

$$\to \quad \phi_a \nabla^2 \phi_a = -\vec{\nabla}\phi_a.\vec{\nabla}\phi_a. \tag{210}$$

The total energy of the system is given by

$$E = \int \frac{1}{2}\vec{\nabla}\phi_a.\vec{\nabla}\phi_a d^2x. \tag{211}$$

We therefore have a S^2 fiber sitting on every point of the xy-plane, forming a bundle space which is locally $S^2 \times R^2$. The classical vacuum of the system is at $\phi_a = constant$. E vanishes for the vacuum. The $O(3)$ symmetry is spontaneously broken by the vacuum, which can be arbitrarily chosen to be at $(0,0,1)$. A finite-energy, localized solution of (209) is described by two functions $\phi_1(x,y)$ and $\phi_2(x,y)$ with $\phi_{1,2} \to 0$ as $r \to \infty$. The points on the xy-plane can therefore be transformed to a sphere using a stereographich map, in which all points on a circle with $r \to \infty$ are mapped to the "north pole". This identification is allowed so long as the single-valuedness of the ϕ_a-field is concerned, since all points on this large circle reside at the same vacuum point $(0,0,1)$. In solid state theory, the ϕ_a-field may describe the order parameter of a 2D ferromagnet. The $S^2 \to S^2$ mapping from the configuration S^2 to the field S^2, therefore allows the following geometrical description of the system:

- Spin waves are sections of the bundle space which is now compactified to $S^2 \times S^2$.
- Spin waves belong to distinct homotopy classes. The appropriate homotopy group is $\pi_2(S^2) = Z$.
- Each mapping can be characterized by a winding number n which counts the Brouwer degree of the map.

This winding number is obtained as

$$Q = \frac{1}{8\pi}\int \epsilon_{ij}\epsilon_{abc}\phi_a\partial_i\phi_b\partial_j\phi_c d^2x$$

$$= \frac{1}{8\pi}\int \epsilon_{ij}\epsilon_{abc}\phi_a \frac{\partial\phi_b}{\partial\theta_l}\frac{\partial\theta_l}{\partial x^i}\frac{\partial\phi_c}{\partial\theta_m}\frac{\partial\theta_m}{\partial x^j}d^2x = \frac{1}{8\pi}\int \epsilon_{lm}\epsilon_{abc}\phi_a\frac{\partial\phi_b}{\partial\theta_l}\frac{\partial\phi_c}{\partial\theta_m}d^2\theta. \tag{212}$$

But $\frac{1}{2}\epsilon_{lm}\epsilon_{abc}\frac{\partial\phi_b}{\partial\theta_l}\frac{\partial\phi_c}{\partial\theta_m}d^2\theta$ is the surface element of the internal S^2. Therefore

$$Q = \frac{1}{4\pi}\int d\vec{S}_{int}.\vec{\phi} = n. \tag{213}$$

Note that $d\vec{S}_{int}$ is parallel to the radius vector $\vec{\phi}$ of the unit sphere in the (ϕ_1, ϕ_2, ϕ_3) space.

This interesting system can also be formulated on the complex plane, using the stereographic projection

$$\omega = \frac{2\phi_1}{1-\phi_3} + i\frac{2\phi_2}{1-\phi_3} \equiv \omega_1 + i\omega_2. \tag{214}$$

The north pole of the ϕ-sphere is projected to $|\phi| \to \infty$. Note that this projection is conformal (i.e. preserves angles).

Consider the self-duality relation

$$^{*}d\omega = -id\omega, \tag{215}$$

which leads to the equations

$$\frac{\partial\omega_1}{\partial x^1} = \pm\frac{\partial\omega_2}{\partial x^2}, \tag{216}$$

and

$$\frac{\partial\omega_1}{\partial x^2} = \mp\frac{\partial\omega_2}{\partial x^1}. \tag{217}$$

Since these are the familiar analyticity conditions, any analytic function $\omega(z)$ or $\omega(z^*)$, is a solution of (215). It can be shown that such solutions in fact minimize the energy functional

$$E = \int \frac{1}{2}\partial_i\phi_a\partial_i\phi_a d^2x = \int \frac{|d\omega/dz|}{(1+\frac{1}{4}|\omega|^2)^2}. \tag{218}$$

The topological charge for these solutions is

$$Q = \pm\frac{1}{4\pi}E. \tag{219}$$

A simple solution with

$$Q = \frac{1}{4\pi}E = n \tag{220}$$

is given by

$$\omega(z) = \frac{(z-z_o)^n}{\lambda^n}, \tag{221}$$

where z_o and λ are constants. Note that Q and E do not depend on λ and z_o which shows that the solutions can be scaled up or down or displaced on the z-plane.

It is seen that the nonlinear $O(3)$ model is quite simple, yet rich in structure. This simple model provides a good insight into the more complicated systems like the Yang-Mills instantons.

13 Instantons in the CP_N Model

Real projective spaces $P_N(R)$ are lines in R^{N+1} which pass through the origin. For example, $P_3(R) = SO(3)$. CP_N is the complex version of P_N.

Consider $N+1$ complex scalar fields ϕ_a ($a = 1, 2, ..., N+1$), with the following Lagrangian density (for more details, see Eichenherr 1978)

$$\mathcal{L} = \partial^\mu \phi_a^* \partial_\mu \phi_a + \phi_a^* (\partial^\mu \phi_a) \phi_b^* \partial_\mu \phi_b, \tag{222}$$

in which summation over μ and a indices is implied. x^μ ($\mu = 1, 2$) is considered to be the (xy) Euclidean plane. The complex fields are subject to the constraint

$$|\phi_1|^2 + ... + |\phi_{N+1}|^2 = 1, \tag{223}$$

or $\phi_a^* \phi_a = 1$. These complex fields form an N-dimensional complex projective (CP_N) space. It is interesting to note that the Lagrangian density (222) is invariant under the local gauge transformation

$$\phi_a'(x) = \phi_a(x) e^{i\Lambda(x)} \tag{224}$$

without a need to introduce any gauge potentials. A vector field $A^\mu(x)$, however, can be defined according to

$$A^\mu = i\phi_a^* \partial^\mu \phi_a, \tag{225}$$

which is real since $\phi_a^* \partial^\mu \phi_a$ is pure imaginary. Using this definition, the Lagrangian density (222) can be written as

$$\mathcal{L} = (D^\mu \phi_a)^* D_\mu \phi_a, \tag{226}$$

where

$$D_\mu \phi_a = (\partial_\mu + iA_\mu) \phi_a, \tag{227}$$

in resemblance to the covariant differentiation in electromagnetism. Note, however, that A^μ is not a new degree of freedom, and it is determined solely in terms of the ϕ_a field. Localized solutions with finite energy should satisfy the following asymptotic behavior

$$\phi_a \to C_a e^{i\psi(\theta)}, \tag{228}$$

where C_a are constants satisfying $C_a^* C_a = 1$, and $\psi(\theta)$ is the common phase of the fields (θ is the polar angle in the xy-plane). This phase implies the winding number

$$Q = \frac{1}{2\pi} \int d\psi = \frac{1}{2\pi} \oint \frac{d\psi}{d\theta} d\theta = n, \tag{229}$$

where n is an integer. Similar to what we had in the case of cosmic strings, a $\pi_1(S^1)$ homotopy group is involved and solutions having different winding numbers belong to distinct homotopy classes.

14 Skyrme Model

T.H.R. Skyrme introduced a nonlinear model in 1960s, with soliton solutions approximately describing baryons (Skyrme, 1961). This model is based on the nonlinear sigma model, augmented with a nonlinear term in the Lagrangian which stabilizes the solitons. The Skyrme Lagrangian is therefore given by

$$\mathcal{L}_{Sk} = -\frac{f_\pi^2}{4} Tr(U^\dagger \partial_\mu U U^\dagger \partial^\mu U) + \frac{1}{32\alpha^2} Tr([U^\dagger \partial_\mu U, U^\dagger \partial_\nu U]^2), \tag{230}$$

where U is the field described by a unitary matrix, α is a dimensionless coupling constant ($\simeq 5$), and f_π^2 is the pion decay constant (130-190 MeV). The pion mass term can also be included in (230) by adding the term $-\frac{1}{2}m_\pi Tr(U + U^\dagger)$. Soliton solutions of the Skyrme model are called *Skyrmions.* They have interesting topological properties, in relevance to the low energy properties of baryons. Witten (1983) showed that the Skyrme model is the high-N_c limit of QCD, where N_c is the number of colors. For a review of the mathematical developments in the Skyrme model, the reader is referred to Gisiger and Paranjape (1998). The geometrical aspects of the Skyrme model were first discussed by Manton and Ruback (1986).

Since U becomes constant at spatial infinity, the points in R^3 can be mapped onto a S^3. Soliton solutions, therefore, define a mapping from this S^3 to the group manifold of $SU(2)$ which is also an S^3. These solutions can therefore be classified according to the third homotopy group π_3:

$$\pi_3 : S^3 \longrightarrow S^3. \tag{231}$$

This homotopy group is isomorphic to the group of integers under addition, and the topological charges label the distinct sectors. The topological current of the Skyrme model is given by

$$B^\mu = \frac{1}{24\pi^2} \epsilon^{\mu\nu\alpha\beta} Tr(U^\dagger \partial_\nu U U^\dagger \partial_\alpha U U^\dagger \partial_\beta U), \tag{232}$$

with the corresponding charge identified with the baryon number:

$$B = \frac{1}{24\pi^2} \int d^3x \epsilon^{ijk} Tr(U^\dagger \partial_i U U^\dagger \partial_j U U^\dagger \partial_k U). \tag{233}$$

The simplest Skyrmion is obtained using the ansatz

$$U = e^{i\hat{r}.\vec{\tau} f(r)}, \tag{234}$$

where $\vec{\tau} = (\tau_1, \tau_2, \tau_3)$ are the Pauli matrices, $\hat{r}$ is the unit vector in the radial direction, and $f(r)$ is a function to be determined by the field equation. The boundary conditions which are required to have a well-defined solution are

$$f(0) = \pi \quad and \quad f(\infty) = 0. \tag{235}$$

The simplest description of the Skyrme model is in terms of unitary 2×2 matrices U which belong to $SU(2)$.

Houghton et al. (1998) introduced a new ansatz for the Skyrme model

$$U(r,z) = exp(if(r)\hat{n}_R.\vec{\sigma}), \tag{236}$$

where

$$\hat{n}_R = \frac{1}{1+|R|^2}(2Re(R), 2Im(R), 1-|R|^2). \tag{237}$$

Note that the complex coordinate z is related to the polar coordinates θ and ϕ via $z = tan(\theta/2)exp(i\phi)$ and $R(z)$ is a rational map $R(z) = p(z)/q(z)$, where p and q are polynomials of maximum degree N. The boundary conditions $f(0) = k\pi$ ($k \in Z$) and $f(\infty) = 0$ are implied. The degree of the rational map $R(z)$ determines the baryon number ($B = Nk$). Therefore, all baryon numbers can be obtained with $k = 1$, using the ansatz (236). The $N = 1$ case is the Skyrme's original hedgehog ansatz. The ansatz (236) leads to the following expression for the total energy (Houghton et al. 1998)

$$E = \int \left[\lambda_1^2 + \lambda_2^2 + \lambda_3^2 + (\lambda_1\lambda_2)^2 + (\lambda_2\lambda_3)^2 + (\lambda_1\lambda_3)^2\right] d^3x, \tag{238}$$

where λ_1^2, λ_2^2, and λ_3^2 are the eigenvalues of the symmetric strain tensor

$$D_{ij} = -\frac{1}{2}Tr((\partial_i U U^{-1})(\partial_j U U^{-1})). \tag{239}$$

The baryon number density is given by

$$b = \frac{1}{2\pi^2}\lambda_1\lambda_2\lambda_3. \tag{240}$$

It can be shown that

$$\lambda_1 = -f'(r), \quad and \quad \lambda_2 = \lambda_3 = \frac{\sin f}{r}\frac{1+|z|^2}{1+|R|^2}|\frac{dR}{dz}|. \tag{241}$$

A Bogomol'nye-type lower limit to the energy functional exists:

$$E \geq 4\pi^2(2N + \sqrt{I}), \tag{242}$$

where

$$I = \frac{1}{4\pi}\int \left(\frac{1+|z|^2}{1+|R|^2}|\frac{dR}{dz}|\right)^4 \frac{2idzd\bar{z}}{(1+|z|^2)^2}. \tag{243}$$

The energy limit can be written as $E \geq 12\pi^2 N$, since I itself satisfies the inequality $I \geq N^2$.

15 Solitons and Noncommutative Geometry

Noncommutative geometry provides a nice tentative framework for the unified description of gauge fields and nonlinear scalar fields responsible for the spontaneous breakdown of gauge symmetries (Connes 1985, Coquereaux et al. 1991, Chamseddine et al., 1993a, Madore 1995, and Okummura et al. 1995).

It has been shown that noncommutative geometry can also deal with gravity and leads to generalized theories of gravity like scalar-tensor theories (Chamseddine et al. 1993b). Here, we briefly describe the general mathematical structure of the Z_2-graded noncommutative geometry and its relevance to the localized soliton-like solutions, closely following Teo and Ting (1997). The reader is referred to this paper for further details.

The Yang-Mills-Higgs theory is formulated using differential forms on $\mathcal{M} \times Z_2$, where $\mathcal{M}$ is an $n-$dimensional Euclidean space and Z_2 is the cyclic group of order two

$$Z_2 = \{e, r | r^2 = e\}. \tag{244}$$

An explicit matrix representation of this group is

$$\pi(e) = \begin{pmatrix} 1 & 0 \\ 0 & 1 \end{pmatrix}, \quad and \quad \pi(r) = \begin{pmatrix} 1 & 0 \\ 0 & -1 \end{pmatrix}. \tag{245}$$

The algebra of complex functions on Z_2 is denoted by $\mathcal{M}_2^+$ and is a subalgebra of $\mathcal{M}_2$ which is the algebra generated by the Pauli matrices. An element of $\mathcal{M}_2^+ \otimes C$ is thus in the following form

$$\begin{pmatrix} f_1 & 0 \\ 0 & f_2 \end{pmatrix}, \tag{246}$$

where f_1 and f_2 are complex functions.

A generalized $p-$form on $\mathcal{M} \times Z_2$ looks like

$$\eta = \begin{pmatrix} A + B & C \\ C' & A' + B' \end{pmatrix}, \tag{247}$$

where A and A' are $p-$forms on $\mathcal{M}$ (the horizontal $p-$forms), B and B' are $(p-2)-$forms on $\mathcal{M}$, and C and C' are $(p-1)-$forms on $\mathcal{M}$. Note that η is a composite $p-$form on $\mathcal{M} \times Z_2$, since a generalized $p-$form can be written as $\eta = a \times A$ where a is a $q-$form on Z_2 (vertical part) and A is a $(p-q)-$form on $\mathcal{M}$ (horizontal part). Even-degree forms on Z_2 are diagonal matrices while those of odd degree are off-diagonal. Operations on differential forms like exterior differentiation and Hodge * operation can be extended to the Z_2 forms.

The generalized connection 1-form on $\mathcal{M} \times Z_2$ has the form

$$\omega = \mathcal{A} + \theta + \phi, \tag{248}$$

where $\mathcal{A}$ is the Yang-Mills connection (horizontal part of ω), and $\theta + \phi$ is the vertical part associated with the internal Z_2. The corresponding internal symmetry consists of two global $U(1)$'s for the two elements of Z_2, denoted by $U(Z_2)$. Under $g \in U(Z_2)$, θ is the gauge invariant Maurer-Cartan 1-form. The curvature 2-form corresponding to the connection (248) is

$$\Omega = d\omega = i\omega \wedge \omega = F + D_H\phi + m^2 - \phi^2, \tag{249}$$

where $F = d_H\mathcal{A} + i\mathcal{A} \wedge \mathcal{A}$ is the Yang-Mills curvature and

$$D_H\phi = d_H\phi + i\mathcal{A} \wedge \phi. \tag{250}$$

Writing $\mathcal{A}$ and ϕ explicitly in their matrix forms

$$\mathcal{A} = \begin{pmatrix} A & 0 \\ 0 & B \end{pmatrix}, \quad \phi = \begin{pmatrix} 0 & \phi \\ \phi^\dagger & 0 \end{pmatrix}, \tag{251}$$

where A and B are ordinary 1-forms and ϕ is a complex scalar field, the curvature 2-form becomes

$$\Omega = \begin{pmatrix} F + m^2 - \phi\phi^\dagger & -D_H\phi \\ -D_H\phi^\dagger & G + m^2 - \phi^\dagger\phi \end{pmatrix}, \tag{252}$$

where $F = d_HA + iA \wedge A$, $G = d_HB + iB \wedge B$, and $D_H\phi = d_H\phi + i(A\phi - \phi B)$. The Euclidean action functional then becomes

$$S = \frac{1}{2}\int d^nxTr(\Omega^\dagger_{ij}\Omega^{ij})$$

$$= \int d^nx \left\{\frac{1}{2}F_{\alpha\beta}F^{\alpha\beta} + \frac{1}{2}G_{\alpha\beta}G^{\alpha\beta} + 2D_\alpha\phi^\dagger D^\alpha\phi + 2(m^2 - \phi^\dagger\phi)^2\right\}. \tag{253}$$

Note that $D\Omega = 0$ (Bianchi identity) and extremization of (253) leads to $D^*\Omega = 0$ which can be expanded into the following equations

$$D_\beta F_{\alpha\beta} = i(D_\alpha\phi^\dagger\phi - \phi^\dagger D_\alpha\phi),$$

$$D_\beta G_{\alpha\beta} = i(D_\alpha\phi^\dagger\phi - \phi^\dagger D_\alpha\phi^\dagger),$$

and

$$D_\alpha D_\alpha\phi = -2(m^2 - \phi\phi^\dagger)\phi. \tag{254}$$

An element of the composite symmetry group has the form $g = diag(g_1, g_2)$ where g_1 and g_2 belong to the Yang-Mills gauge group. In the special case $g_1 = g_2$ and $A = B$, the action functional reduces to

$$S = \int d^n x \left\{ \frac{1}{4} F_{\alpha\beta} F^{\alpha\beta} + \frac{1}{2} D_\alpha \phi^\dagger D^\alpha \phi + \frac{\lambda}{2} (\phi^\dagger \phi)^2 - \mu^2 \phi^\dagger \phi \right\} \tag{255}$$

where $\mu = em$, $\lambda = e^2$ and the rescaling $\omega \to e\omega$ has been implemented. If the gauge group responsible for the connection A is $SU(2)$, this action will represent the ordinary Euclidean Yang-Mills theory coupled to a Higgs field with $F_{\alpha\beta} = \partial_\alpha A_\beta - \partial_\beta A_\alpha + ie[A_\alpha, A_\beta]$ and $D_\alpha \phi = \partial_\alpha \phi + ie[A_\alpha, \phi]$. The interesting aspect of the above formalism is that the Higgs field ϕ responsible for the spontaneous symmetry breaking of the Yang-Mills gauge field is now a gauge field associated with the internal Z_2 group. Moreover, the quadratic potential $V(\phi, \phi^\dagger)$ results naturally from the curvature of the connection $\omega = \theta + \phi$ (i.e. $\Omega = d\omega + i\omega \wedge \omega = m^2 - \phi^2$, where only the vertical part has now been considered).

If the manifold $\mathcal{M}$ is one dimensional, the horizontal guage field A dissappears since A_μ has only one component $\mu = 1$ and the corresponding curvature vanishes. The Lagrangian density then reduces to

$$\mathcal{L} = \frac{1}{2} \left| \frac{d\phi}{dx} \right|^2 + \frac{e^2}{2} (|\phi|^2 - m^2)^2, \tag{256}$$

which leads to the field equation

$$\frac{d^2\phi}{dx^2} = 2e^2 \phi (|\phi|^2 - m^2). \tag{257}$$

The ansatz $\phi(x) = \chi(x) exp(ikx)$ with real $\chi(x)$, leads to $k = 0$ and

$$\phi(x) = \chi(x) = \pm m \tanh(emx). \tag{258}$$

This is the familiar kink (anti-kink) solution of the nonlinear Klein-Gordon equation in the static case.

The topological current $J = \frac{1}{2m} \frac{\partial \phi}{\partial x}$ leads to the charge $Q = \int J dx = \pm 1$ for the kink (anti-kink). The kink (anti-kink) solution satisfies the generalized (anti-) self-duality relations

$$^*\Omega \wedge \theta = \pm im\Omega. \tag{259}$$

This self-duality condition also provides a lower limit for the soliton energy. This lower limit ($E = \frac{4}{3} em^3$) is realized for the kink (anti-kink) solution.

In a 2-dimensional Euclidean space with $U(1)$ as the underlying gauge symmetry (255) reduces to the static abelian Higgs model (see equation 162). The string solution of Nielsen and Olesen fulfills the (anti-) self-duality relation

$$^*\Omega = \pm \Omega, \tag{260}$$

similar to that of the non-abelian magnetic monopole. Note that here, $\Omega_{23} = -\frac{1}{m}D_2(Re\phi + iIm\phi\tau_3)$, and $\Omega_{34} = \frac{e}{m^2}(m^2 - \phi^*\phi)\tau_3$ where $B = F_{12}$ is the magnetic field, τ_3 is the third Pauli matrix, and D_α is the $U(1)$ covariant derivative.

In the $n = 3$ case with $G = SU(2)$, the action (255) can be identified with the static Yang-Mills-Higgs system which possesses the monopole solutions of 't Hooft and Polyakov. The (anti) self-duality conditions (260) are now satisfied by the exact solutions of Prasad and Sommerfield in the $\lambda \to 0$ limit.

Acknowledgements

I would like to thank the Physics and Astronomy Department of the University of Victoria in B.C. and in particular F. Cooperstock for hospitality during my sabbatical leave. Helpful suggestions from V. Karimipur and M. Hakim-Hashemi is gratefully acknowledged. I also thank the Shiraz University Research Council for finantial support.

References

[1] Belavin, A.A., and Polyakov, A.M., *JETP Lett.*, **22**, 245, 1975.

[2] Bogomol'nyi, E.B., *Sov. J. Nucl. Phys.*, **24**, 449, 1976.

[3] Chamseddine, A.H., Felder, G., and Fröhlich, J., *Nucl. Phys.*, **B395**, No. 3, 672, 1993a.

[4] Chamseddine, A.H., Felder, G., and Fröhlich, J., *Comm. Math. Phys.*, **155**, 205, 1993b.

[5] Chern, S.S., *Complex Manifolds Without Potential Theory*, D. Van Nostrand Co., 1967.

[6] Coleman, S., *Classical Lumps and their Quantum Descendents*, Erice Lecture (1975), in New Phenomena in Subnuclear Physics, Ed. A. Zichichi, Plenum, N.Y., 1977.

[7] Connes, A., *Non-Commutative Differential Geometry*, *Publ. I.H.E.S.*, **62**, 41, 1985.

[8] Coquereaux, R., Esposito-Farese, G., and Vaillant, G., *Nucl. Phys.*, **B353**, 689, 1991.

[9] Coquereaux, R., Esposito-Farese, G., and Scheck, F., *Int. J. Mod. Phys.*, **A7**, No. 26, 6555, 1992.

[10] Darling, P.W.R., *Differential Forms and Connections*, Cambridge University Press, USA, 1994.

[11] Dijkgraaf, R., Witten, E., *Comm. Math. Phys.*, **129**, 393, 1990.

[12] Drechsler, W., and Mayer, M.E., *Fiber Bundle Techniques in Gauge Theories*, Lecture notes in physics 67, Springer Verlag, Heidelberg, 1977.

[13] Duan, Y., Xu, T., and Fu, L., *Prog. Theor. Phys.*, **101**, No. 2, 467, 1999.

[14] Eguchi, T., Gilkey, P., and Hanson, A.J., *Phys. Rep.*, **66**, No. 6, 213, 1980.

[15] Eichenherr, H., *Nucl. Phys.*, **B146**, 215, 1978.

[16] Felsager, B., *Geometry, Particles, and Fields (2nd Ed.)*, Odense University Press., Gylling, 1983.

[17] Gisiger, T., and Paranjape, M.B., *Phys. Rep.* , **306**, 109, 1998.

[18] Goddard, P., and Olive, D., *Magnetic Monopoles in Gauge Field Theories, Reports in Progress in Physics*, **41**, 1357, 1978.

[19] Guillemin, V., and Pollack, A., *Differential Topology*, Prentice-Hall Inc., USA, 1974.

[20] Henneaux, M., *Contemp. Math.*, **219**, (AMS), 1998.

[21] Horvathy, P.A., *Introduction to Monopoles*, Bibliopolis, Napoli, 1988.

[22] Houghton, C.J., Manton, N.S., and Sutcliffe, P.M., *Nucl. Phys.*, **B510**, No. 3, 507, 1998.

[23] Henneaux, M., et al., *Comm. Math. Phys.*, **186**, 137, 1997.

[24] Lamb, G.L., *Elements of Soliton Theory*, John Wiley and Sons, New York, 1980.

[25] Liu, F., and Mazenko, G.F., *Phys. Rev. Lett.* , **78**, 401, 1997.

[26] Long, P.E., *An Introduction to General Topology*, Charles E. Merrill Pub. Co., U.S.A., 1971.

[27] Madore, J., *An Introduction to Noncommutative Geometry and its Applications*, Cambridge University Press, Cambridge, England, 1995.

[28] Manton, N.S., and Ruback, P.J., *Phys. Lett.*, **B181**, 137, 1986.

[29] Morris, M.S., and Thorne, K.S., *Am. J. Phys.*, **56**, 395, 1988.

[30] Nambu, Y., *Lectures in Applied Math*, **21**, (AMS), 1985.

[31] Nielsen, N.K., and Olesen, P., *Nucl. Phys.* , **B61**, 45, 1973.

[32] Nielsen, N.K., and Olesen, P., *Nucl. Phys.* , **B291**,No. 4, 829, 1987.

[33] Okummura et al., *Prog. Theor. Phys.* , **94**, 1121, 1995.

[34] Polyakov, A.M., *JETP Lett.*, **20**, 194, 1974.

[35] Prasad, M.K., and Sommerfield, C.H., *Phys. Rev. Lett.*, **35**, 760, 1975.

[36] Preskill, J., *Vortices and Monopoles*, Les Houches Lectures (1985), Caltech preprint CALT-86-1287.

[37] Rajaraman, R., *Solitons and Instantons*, Elsevier, The Netherlands, 1987.

[38] Riazi, N., and Gharaati, A.R., *Int. J. Theor. Phys.*, **37**, No. 3, 1081, 1998.

[39] Singer, I.M., and Thorpe, H.A., *Lecture Notes in Elementary Geometry and Topology*, Glenview, I11, Scott, Foresman, 1967.

[40] Skyrme, T.H.R, *Proc. Roy. Soc. London*, **260**, 127, 1961.

[41] Teo, E., and Ting, C., *Phys. Rev.* , **D56**, No. 4, 2291, 1997.

[42] T'Hooft, G., *Nucl. Phys.*, **B79**, 276, 1974.

[43] Vasheghani, A., and Riazi, N., *Int. J. Theor. Phys.*, **35**, 587, 1996.

[44] Yang, C.N., *Phys. Rev. Lett.* , **33**, No. 7, 445, 1974.

[45] Yang, C.N., and Mills, R.L., *Phys. Rev.* , **96**, 191, 1954.

[46] Witten, E., *Nucl. Phys.*, **B223**, 433, 1983.

[47] Wu, T.T., and Yang, C.N., *Phys. Rev.*, **D12**, 3845, 1975.

Theoretical Physics 2002, Part 1
Edited by Thomas F. George and Henk F. Arnoldus, pp. 181-214

ISBN 1-59033-435-3

Rotations of Perturbed Triaxial Rigid Bodies

Antonio Elipe*

Abstract

Attitude dynamics of perturbed triaxial rigid bodies is a rather involved problem, due to the presence of elliptic functions even in the Euler equations for the free rotation of a triaxial rigid body.

We have revisited the Euler-Poinsot problem, and present several sets of canonical variables as well as its integration. With the solution of this problem, that will be taken as the unperturbed part, we expand the perturbation in Fourier series, which coefficients are polynomials of the Jacobian nome. These series converge very fast, and thus, with only few terms a good approximation is obtained. Once the expansion is performed, it is possible to apply to it a Lie-transformation. Several applications to artificial and natural bodies have been carried out.

With the solution of the unperturbed problem, other aspects, like transversal homoclinic orbits and stochasticity bands are also considered.

1 Introduction

Attitude dynamics of rigid bodies is a common problem in both natural and artificial celestial bodies, and has been the object of study since the beginning of the celestial mechanics. Indeed, much of the work done in this regard is associated with names of great scientists like Euler, Hamilton, Poinsot, Lagrange, etc. Their work still is valid today, either in the aspect of attitude representation or in attitude dynamics [1].

Space Age brought to astronomers, among other things, the fact that, apart from the classical bodies like the planets of satellites that are almost spherical, the shape of many natural celestial bodies is far away from the sphere; they present a very irregular shape, which adds a lot of complexity to the motion, that in some cases has been proved to be chaotic [2]. Besides, spacecrafts are also built under several models, and their antennas modify the distribution of masses, and even introduce a big component of elasticity.

Albeit most celestial bodies of interest are not perfect rigid bodies, usually, the rigid body model is a good approximation for describing the rotational motion, and hence the interest in having good theories for perfect rigid bodies.

*Grupo de Mecánica Espacial, Universidad de Zaragoza. 50009 Zaragoza. Spain

The simplest case we meet in the study of rigid body motion is the so called *Euler-Poinsot* problem, that consists of describing the motion of a free rigid body in the space. The attitude is described by elements of the group of rotations SO(3). The Hamiltonian is essentially the kinetic energy, defined on the cotangent space. Since there are no external forces, the angular moment vector is an integral, and also are its components and its norm. Thus, the problem is integrable.

Of capital importance when integrating or analyzing perturbed problems is the choice of a convenient set of variables. Depending on the type of studies to be carried out, one set of variables may fit the problem better than others. With this aim, we present in this work several types of solutions for the Euler-Poinsot problem. Firstly, we obtain the solution in terms of the body components of the angular moment vector, that lie on a sphere (Section 2). By studying the rotational motion in these variables, one finds the phase flow and types of motion, which will be useful to find new sets of variables.

The first canonical variables we meet are based on the Euler angles (Section 3). Since they reflects the SO(2) symmetry of rotations about the third space axis, the problem is reduced to a two degrees of freedom problem. But the system is also symmetric with respect to the group SO(3), which makes it possible to reduce the system to only one degree of freedom. This is accomplished with the Serret-Andoyer variables. A complete reduction may be obtained by means of the transformations given by Sadov [3] and Kinoshita [4], that give similar sets of action and angle variables. Finally, in this section, another transformation [5] that converts the Hamiltonian into a function depending only on the momenta is presented.

It is well known that three angles cannot afford a regular representation of the group SO(3). A solution was proposed by Euler himself and consists of the use of the so called *Euler parameters*, a set of four quantities. Rotations, thus, can be handled as quaternions of unit length (Section 4). For the components of a unit quaternions describing a rotation, we find their conjugate moments, and the Hamilton equations become very simple. Since quaternions algebra only involves elementary operations, with no need to compute trigonometric functions, quaternions are good candidates for numerical integration. For the Serret-Andoyer variables we do the same [5], but now, since the problem in these variables has only one freedom, we handle the dynamics by means of only one quaternion, in which two of its components are coordinates and the remaining two components are their canonically conjugate moments.

Once made the detailed treatment of the free motion, we continue by considering several problems. The first one we take into consideration is the motion of a tri-axial rigid body under a gravity field (Section 5) that may be used to describe the attitude of a spacecraft moving around a spherical planet, or to describe the Earth precession [6], for instance. Here, we use Sadov's solution of the rigid body in free rotation (the unperturbed part), and expand the potential (the perturbation) as a Fourier series of the Jacobian nome. These expansions are very fast convergent and only few terms give enough accuracy. The expansions are carried out automatically by our own software [7] with the technique described in [8]. An application to a tri-axial body in a Newtonian gravity field is made.

In regard with qualitative aspects, those variables that give a geometrical insight of the problem are usually preferred. As an illustration, we consider the rigid body in free rotation with a periodic perturbation (Section 6). The components of the angular momentum give a clear picture of the phase flow, and on the sphere we can represent the six equilibria. By means of the solution for the heteroclinic trajectory given in Section 2 we compute the Melnikov function [9], and show that in this case, the perturbation give rises to chaotic motion in the sense that the system has Smale's horseshoes.

Lastly (Section 7), we consider an extension of the rigid body, a gyrostat or dual-spin body in free rotation. For such a body, the angular moment is an integral, but the Hamiltonian is not made of the energy. The system has the same symplectic structure as the rigid body, and the problem is shown to be equivalent to a quadratic Hamiltonian on the unit sphere. At difference with the rigid body that has six equilibria, here, the number of equilibria depends on the shape of the body and also on the number of rotors, which presents a rich variety of bifurcations. Again, the components of the angular moment reveals to be of great utility for handling this problem, since in other representations, like Serret-Andoyer variables, some points cannot be represented [5] and these points, precisely, are good candidates to bifurcate.

2 Components of the Angular Momentum as Variables

To study the attitude of a rigid body about a point C, we set two right-handed frames at that point. One fixed in the space $\mathcal{S} \equiv C\boldsymbol{s}_1\boldsymbol{s}_2\boldsymbol{s}_3$, and another one $\mathcal{B} \equiv C\boldsymbol{b}_1\boldsymbol{b}_2\boldsymbol{b}_3$, fixed in the body. As it known from Euler, attitude of $\mathcal{B}$ in $\mathcal{S}$ results from three rotations. Among the several sequences of rotations (see [1]) we choose the 3-1-3, that will be explained below.

Firstly, we define the unit vector $\boldsymbol{\ell} = \boldsymbol{s}_3 \times \boldsymbol{b}_3 \sin\vartheta$, with $0 \leq \vartheta \leq \pi$ the *inclination* angle or nutation angle ($\cos\vartheta = \boldsymbol{s}_3 \cdot \boldsymbol{b}_3$). This vector lies on the intersection line of the planes $C\boldsymbol{s}_1\boldsymbol{s}_2$ and $C\boldsymbol{b}_1\boldsymbol{b}_2$, thus, there exists an angle ϕ with $0 \leq \phi \leq 2\pi$, called *precession* angle, such that

$$\boldsymbol{\ell} = \boldsymbol{s}_1 \cos\phi + \boldsymbol{s}_2 \sin\phi.$$

But $\boldsymbol{\ell}$ is also on the plane $C\boldsymbol{b}_1\boldsymbol{b}_2$, and consequently is a linear combination of these vectors

$$\boldsymbol{\ell} = \boldsymbol{b}_1 \cos\psi - \boldsymbol{b}_2 \sin\psi,$$

where the angle ψ ($0 \leq \psi \leq 2\pi$) is the longitude of the body vector $\boldsymbol{b}_1$ reckoned from the node $\boldsymbol{\ell}$, and is usually known as the *proper rotation*.

With these definitions, the sequence 3-1-3 above mentioned is the rotation

$$R = R(\phi, \boldsymbol{s}_3) \circ R(\vartheta, \boldsymbol{\ell}) \circ R(\psi, \boldsymbol{b}_3), \tag{1}$$

and by applying Rodrigues' Theorem for transposition of rotations [10], we find the differential of the rotation as

$$dR = \boldsymbol{s}_3 d\phi + \boldsymbol{\ell} d\vartheta + \boldsymbol{b}_3 d\psi. \tag{2}$$

Since this differential is expressed in terms of the non-orthonormal basis $(\boldsymbol{s}_3, \boldsymbol{\ell}, \boldsymbol{b}_3)$, we can express the rest of the vectors in such basis; thus, for example, we have

$$\begin{aligned} \boldsymbol{b}_1 &= \boldsymbol{\ell} \cos\psi + \frac{\boldsymbol{s}_3 - \boldsymbol{b}_3 \cos\vartheta}{\sin\vartheta} \sin\psi, \\ \boldsymbol{b}_2 &= -\boldsymbol{\ell} \sin\psi + \frac{\boldsymbol{s}_3 - \boldsymbol{b}_3 \cos\vartheta}{\sin\vartheta} \cos\psi, \end{aligned} \tag{3}$$

that will be of utility later on.

As it was stated in [5], the angular momentum vector $\boldsymbol{G}$ of the body serves to overlay a symplectic structure on the phase space of the Euler-Poinsot problem through the requirement that the moments Φ, Θ and Ψ, conjugate moments of the Euler angles, satisfy the Cartan 1-form

$$\boldsymbol{G} \cdot dR = \Phi\, d\phi + \Theta\, d\vartheta + \Psi\, d\psi.$$

By means of the expression (2), one easily obtains that

$$\Phi = \boldsymbol{G} \cdot \boldsymbol{s}_3, \qquad \Theta = \boldsymbol{G} \cdot \boldsymbol{\ell}, \qquad \Psi = \boldsymbol{G} \cdot \boldsymbol{b}_3, \tag{4}$$

that is, the conjugate moments are precisely the projections of the angular moment on the non orthonormal basis above defined.

Let us denote by g_1, g_2 and g_3 the components of the angular momentum vector $\boldsymbol{G}$ in the body frame, i.e. $\boldsymbol{G} = g_1 \boldsymbol{b}_1 + g_2 \boldsymbol{b}_2 + g_3 \boldsymbol{b}_3$. By means of Eqs. (4) and (3), it is possible to obtain these components in terms of the canonical Euler variables as

$$\begin{aligned} g_1 &= \left(\frac{\Phi - \Psi \cos\vartheta}{\sin\vartheta} \right) \sin\psi + \Theta \cos\psi, \\ g_2 &= \left(\frac{\Phi - \Psi \cos\vartheta}{\sin\vartheta} \right) \cos\psi - \Theta \sin\psi, \\ g_3 &= \Psi. \end{aligned} \tag{5}$$

It is a matter of computing partial derivatives to check that the Poisson brackets are

$$(g_1; g_2) = -g_3, \qquad (g_2; g_3) = -g_1, \qquad (g_3; g_1) = -g_2. \tag{6}$$

Assuming the body frame coincides with the principal axes of inertia system, the Hamiltonian of the Euler-Poinsot problem is

$$\mathcal{H} = \tfrac{1}{2}(a_1 g_1^2 + a_2 g_2^2 + a_3 g_3^2), \tag{7}$$

with $a_i = 1/I_1$, and I_i the principal moments of inertia. Without loss of generality, we assume that $a_1 > a_2 > a_3$.

Taking into account the structural identities (6), we obtain the Eulerian equations

$$\begin{aligned} \dot{g}_1 &= (g_1, \mathcal{H}) = -(a_2 - a_3) g_2 g_3, \\ \dot{g}_2 &= (g_2, \mathcal{H}) = (a_1 - a_3) g_1 g_3, \\ \dot{g}_3 &= (g_3, \mathcal{H}) = -(a_1 - a_2) g_1 g_2 \end{aligned} \tag{8}$$

describing the motion of $\boldsymbol{G}$ in the body frame.

This system is integrable because it admits two integrals, the energy (7) (7) and the norm of the angular momentum $\|\boldsymbol{G}\| = G$. Indeed, from the equations of motion (8) there results

$$g_1 \dot{g}_1 + g_2 \dot{g}_2 + g_3 \dot{g}_3 = 0,$$

hence

$$g_1^2 + g_2^2 + g_3^2 = \|\boldsymbol{G}\|^2 = G^2 = \text{constant}. \tag{9}$$

N.B. This result is also obtained from the fact that as in the Euler-Poinsot problem there are not external forces, the angular momentum vector is an integral, and hence is its norm, quantity that is independent of the reference frame.

With this integral, we may regard the phase space of (8) as a foliation of invariants manifolds

$$S^2(G) = \{(g_1, g_2, g_3) | g_1^2 + g_2^2 + g_3^2 = G^2\}. \tag{10}$$

The trajectories will be the level contours of the ellipsoid (7) on the sphere (9), as can be seen in Figure 1.

In most of the textbooks, the equations of motion are given in terms of the components of the angular velocity. By doing so, the trajectories are found as the intersections of two ellipsoids, the one corresponding to the energy, the other to the angular momentum, which depends on the moments of inertia. However, by choosing the components of the angular moment, the geometric model depicting the rotations of the angular momentum vector is a sphere of constant radius, independent of the moments of inertia.

The system (8) has six equilibria, $E_{1,3}$, $E_{2,4}$, $E_{0,5}$ that are the intersections of the sphere (10) with the co-ordinate axes $\pm\boldsymbol{b}_1$, $\pm\boldsymbol{b}_2$, and $\pm\boldsymbol{b}_3$ respectively. It is easy to check that the maximum value of the energy is reached at $E_{1,3}$ and the minimum at $E_{0,5}$. The points $E_{1,3}$ and $E_{0,5}$ are centers, whereas $E_{2,4}$ are saddle points.

By introducing as in [5] a state function

$$\mathcal{A} = 2\mathcal{H}/G^2, \tag{11}$$

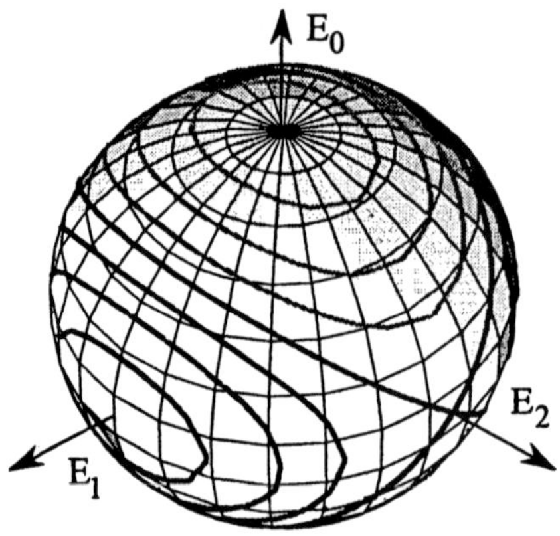

Figure 1: Phase flow of the Euler-Poinsot problem

which inverse has the dimension of a moment of inertia, and assuming that $G \neq 0$ there follows that

$$0 < a_3 \leq \mathcal{A} \leq a_1,$$

and the solutions of the system (8) fall into three types (see figure 1), depending on whether $\mathcal{A} > a_2$ (circulation about $\boldsymbol{b}_1$), $\mathcal{A} = a_2$ (motion on the heteroclinic orbit), or $a_2 < \mathcal{A}$ (circulation about $\boldsymbol{b}_3$). Since the phase flow is different in each region, it is necessary to integrate each region separately.

In the first case, $a_1 > \mathcal{A} > a_2 > a_3$. For the integration, we use the two integrals, the energy (7) and the angular momentum (9) to eliminate g_2 and g_3 in terms of g_1 and the integrals. By replacing them into the first of the equations of the motion (8), there results

$$\dot{g}_1^2 = (n_1^0/G)^2(\Gamma_1^2 - g_1^2)(g_1^2 - \gamma_1^2), \tag{12}$$

where the positive functions Γ_1 and γ_1 are defined by

$$(a_1 - a_3)\Gamma_1^2 = (\mathcal{A} - a_3)G^2, \qquad (a_1 - a_2)\gamma_1^2 = (\mathcal{A} - a_2)G^2,$$

and

$$n_1^0 = G\sqrt{(a_1 - a_2)(a_1 - a_3)}.$$

With this, equation (12) may be reduced to its standard Legendre form, and thus we obtain (for details, the reader is addressed to [5])

$$\begin{aligned} g_1^2 &= G^2 \frac{\mathcal{A} - a_3}{a_1 - a_3} \,\mathrm{dn}^2(n_1 t, k_1), \\ g_2^2 &= G^2 \frac{a_1 - \mathcal{A}}{a_1 - a_2} \,\mathrm{sn}^2(n_1 t, k_1), \\ g_3^2 &= G^2 \frac{a_1 - \mathcal{A}}{a_1 - a_3} \,\mathrm{cn}^2(n_1 t, k_1), \end{aligned} \tag{13}$$

with the modulus $0 < k_1 < 1$ and frequency n_1 such that

$$k_1^2 = \frac{(a_1 - \mathcal{A})(a_2 - a_3)}{(a_1 - a_2)(\mathcal{A} - a_3)}, \qquad n_1 = G\sqrt{(a_1 - a_2)(\mathcal{A} - a_3)}.$$

For the heteroclinic orbit ($\mathcal{A} = a_2$), if we use the above solution, the modulus is $k_1 = 1$. But taking into account the relations

$$\operatorname{sn}(u, 1) = \tanh u, \qquad \operatorname{cn}(u, 1) = \operatorname{dn}(u, 1) = \operatorname{sech}(u),$$

the solution (13) is converted into

$$\begin{aligned} g_1^2 &= G^2 \frac{a_2 - a_3}{a_1 - a_3} \operatorname{sech}^2(n_2 t), \\ g_2^2 &= G^2 \tanh^2(n_2 t), \\ g_3^2 &= G^2 \frac{a_1 - a_2}{a_1 - a_3} \operatorname{sech}^2(n_2 t), \end{aligned} \tag{14}$$

with the frequency $n_2 = G\sqrt{(a_1 - a_2)(a_2 - a_3)}$.

The obtaining of the third region $\mathcal{A} > a_2$, presents no difficulty it we apply the *reciprocal modulus transformation* [11] to the equations (13), and the result may be consulted in [5].

3 Canonical Variables

3.1 Euler variables

So far, we have a set of canonical variables $(\phi, \vartheta, \psi, \Phi, \Theta, \Psi)$, based on the Euler angles. With the results given in the preceding section, it is very simple to obtain the Hamiltonian in those variables. Indeed, by replacing expressions (5) into (7), the Hamiltonian reads

$$\begin{aligned} \mathcal{H} = &\frac{1}{2}(a_1 \sin^2\psi + a_2 \cos^2\psi)\left(\frac{\Phi - \Psi\cos\vartheta}{\sin\vartheta}\right)^2 + \frac{1}{2}a_3\Psi^2 \\ &+ \frac{1}{2}(a_1 \cos^2\psi + a_2 \sin^2\psi)\Theta^2 + (a_1 - a_2)\left(\frac{\Phi - \Psi\cos\vartheta}{\sin\vartheta}\right)\Theta\sin\psi\cos\psi. \end{aligned} \tag{15}$$

In this expression, the angle ϕ is cyclic, thus, expressed in Euler variables, the problem is a two-degree of freedom problem. But we know that it is integrable, then there must be another set of canonical variables that reduces the degrees of freedom. These variables are the so called Serret-Andoyer variables. These variables, although introduced by Serret [12], are known with different names, depending on the source where they are found by researchers. They are widely known as Andoyer variables because they appear in Andoyer's textbook [13]; they receive also the name of Deprit's variables because a transformation given by Deprit [14]. From our part we decided to assign them the name of his creator followed by their most common name, thus the name of *Serret-Andoyer*.

3.2 Serret-Andoyer variables

The main feature of the Serret-Andoyer variables is the use of an intermediate plane, namely the plane perpendicular to the angular momentum. With this plane, the rotation R mapping the frame $\mathcal{S}$ onto $\mathcal{B}$ is decomposed into the product of five rotations. Let us describe briefly these rotations. Firstly, let us denote by $\boldsymbol{k}$ the unit vector in the direction of the angular momentum, that is to say, $\boldsymbol{G} = G\,\boldsymbol{k}$, and let σ denote the angle such that

$$\left.\begin{aligned} \boldsymbol{s}_3 \cdot \boldsymbol{k} &= \cos\sigma, \\ \boldsymbol{s}_3 \times \boldsymbol{k} &= \boldsymbol{i}\,\sin\sigma, \end{aligned}\right\} \quad \text{with} \quad 0 \leq \sigma \leq \pi.$$

The unit vector $\boldsymbol{i}$ is the node of the plane perpendicular to $\boldsymbol{k}$ in the space plane $(\boldsymbol{s}_1, \boldsymbol{s}_2)$, thus,

$$\boldsymbol{i} = \boldsymbol{s}_1 \cos h + \boldsymbol{s}_2 \sin h, \qquad \text{with} \qquad 0 \leq h \leq 2\pi.$$

Let the angle β such that

$$\left.\begin{aligned} \boldsymbol{k} \cdot \boldsymbol{b}_3 &= \cos\beta, \\ \boldsymbol{k} \times \boldsymbol{b}_3 &= \boldsymbol{j}\,\sin\beta, \end{aligned}\right\} \quad \text{with} \quad 0 \leq \sigma \leq \pi, \qquad \text{and} \quad \|j\| = 1.$$

Then, on the one hand, the vector $\boldsymbol{j}$ is perpendicular to the vector $\boldsymbol{k}$, thus it may be expressed as

$$\boldsymbol{j} = \boldsymbol{i} \cos g + (\boldsymbol{k} \times \boldsymbol{i}) \sin g, \qquad \text{with} \qquad 0 \leq g \leq 2\pi,$$

but on the other hand, it is perpendicular to $\boldsymbol{b}_3$, then

$$\boldsymbol{j} = \boldsymbol{b}_1 \cos \ell - \boldsymbol{b}_2 \sin \ell, \qquad \text{with} \qquad 0 \leq \ell \leq 2\pi.$$

Obviously, the rotation R is

$$R = R(\ell, \boldsymbol{b}_3) \circ R(\beta, \boldsymbol{j}) \circ R(g, \boldsymbol{k}) \circ R(\sigma, \boldsymbol{i}) \circ R(h, \boldsymbol{s}_3),$$

and its differential is

$$dR = \boldsymbol{s}_3\, dh + \boldsymbol{i}\, d\sigma + \boldsymbol{k}\, dg + \boldsymbol{j}\, d\beta + \boldsymbol{b}_3\, d\ell. \tag{16}$$

By proceeding in similar way as we did for the Eulerian variables, and taking into account that the vector $\boldsymbol{k}$ is orthogonal to both $\boldsymbol{i}$ and $\boldsymbol{j}$, and that $\boldsymbol{k}$ has the same direction as the vector $\boldsymbol{G}$, the Cartan 1-form is

$$\boldsymbol{G} \cdot dR = (\boldsymbol{G} \cdot \boldsymbol{s}_3) dh + (\boldsymbol{G} \cdot \boldsymbol{k}) dg + (\boldsymbol{G} \cdot \boldsymbol{b}_3) d\ell = H\, dh + G\, dg + L\, d\ell.$$

Thus, the three angles (ℓ, g, h), complemented with the moments

$$L = \boldsymbol{G} \cdot \boldsymbol{b}_3, \qquad G = \|\boldsymbol{G}\|, \qquad H = \boldsymbol{G} \cdot \boldsymbol{s}_3,$$

define a canonical transformation, more precisely, a Mathieu transformation

$$(\phi, \vartheta, \psi, \Phi, \Theta, \Psi) \longmapsto (\ell, g, h, L, G, H).$$

From the geometry, we easily find that the components of the angular momentum are

$$\begin{aligned} g_1 &= \boldsymbol{G} \cdot \boldsymbol{b}_1 = G \sin\beta \sin\ell = \sqrt{G^2 - L^2} \sin\ell, \\ g_2 &= \boldsymbol{G} \cdot \boldsymbol{b}_2 = G \sin\beta \cos\ell = \sqrt{G^2 - L^2} \cos\ell, \\ g_3 &= \boldsymbol{G} \cdot \boldsymbol{b}_3 = L, \end{aligned}$$

and the Hamiltonian (7) is simply

$$\mathcal{H} = \frac{1}{2} \left[(a_1 \sin^2 \ell + a_2 \cos^2 \ell)(G^2 - L^2) + a_3 L^2 \right]. \tag{17}$$

This Hamiltonian is invariant with respect to the group SO(3) of rotations; indeed, the angles g and h are ignorable, then, the problem is reduced to only one degree of freedom. Note that when the rigid body has axial symmetry ($a_1 = a_2$), the three angles are ignorable.

As we already mentioned, Deprit [14] obtained the phase flow of this Hamiltonian. The motion in the phase space (ℓ, L), a cylinder (see Fig. 2), consists of three different types, namely, libration around the stable points, asymptotic motion along the heteroclinic orbits connecting the unstable points, and circulation, such at we can expect from the analysis made in the previous section. As a matter of fact, these variables corresponds to the region $a_1 > a_2 > \mathcal{A} > a_3$. Albeit these variables are very useful, it is not pleasant to handle a rotation about an axis in the form of libration around an equilibrium. To avoid it, Deprit and Elipe [5] defined a second set of Serret-Andoyer variables, that are convenient for the other region of motion, $a_1 > \mathcal{A} > a_2 > a_3$.

Let us remark that the Hamilton equations corresponding to the Hamiltonian (17) have three equilibria, the points $(0, 0)$ (that corresponds to the point E_2), $(\pi/2, 0)$ (that corresponds to the point E_1), and $(\pi, 0)$ (that corresponds to the point E_4). By symmetry, we recover the point E_3, but the points E_0, E_5 cannot be represented in this set of variables (remember that the phase space with the components of the angular momentum is made of spheres, whereas in Serret-Andoyer variables is a cylinder).

3.3 Sadov and Kinoshita variables

In an independent way, Sadov [3] and Kinoshita [4] obtained sets of action -and angle- variables for the rotational motion of a triaxial rigid body. Both sets of variables are very similar. Essentially, they are obtained, firstly, by solving the Hamilton-Jacobi equation from the Hamiltonian (17), then, by obtaining the action -and angle- variables and the generating function of such transformation. Sadov's transformation is given in terms of the Legendre elliptic functions of the first and

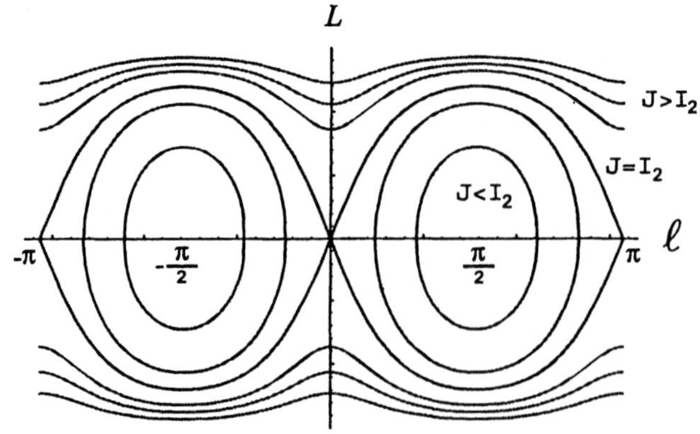

Figure 2: Phase flow of the Euler-Poinsot problem in Serret-Andoyer variables

third kind, whereas in the Kinoshita one, the Heuman Lambda function appears. Both transformation are equivalent [15]. Since, probably, Kinoshita's variables are more popular among journals written in English, we present here Sadov's ones.

The first step consists in obtaining the generating function $\mathcal{S}$ of the transformation

$$(\ell, g, h, L, G, H) \longrightarrow (\beta_1, \beta_2, \beta_3, \alpha_1, \alpha_2, \alpha_3),$$

solution of the Hamilton-Jacobi equation

$$\frac{1}{2}(a_1 \sin^2 \ell + a_2 \cos^2 \ell)\left(\left(\frac{\partial \mathcal{S}}{\partial g}\right)^2 - \left(\frac{\partial \mathcal{S}}{\partial \ell}\right)^2\right) + \frac{1}{2}a_3\left(\frac{\partial \mathcal{S}}{\partial \ell}\right)^2 + \frac{\partial \mathcal{S}}{\partial t} = 0.$$

Since the system is autonomous and the variables g and h are cyclic, the generating function may be put as

$$\mathcal{S} = -\alpha_1 t + \alpha_2 g + \alpha_3 h + \mathcal{U}(\ell; \alpha_1, \alpha_2, \alpha_3),$$

where the function $\mathcal{U}$ is obtained by a quadrature from

$$\left(\frac{\partial \mathcal{U}}{\partial \ell}\right)^2 = I_3\left(\frac{a + b\sin^2 \ell}{c + d\sin^2 \ell}\right),$$

with

$$a = I_1(\alpha_2^2 - 2\alpha_1 I_2), \quad c = I_1(I_3 - I_2) > 0,$$
$$b = (I_2 - I_1)\alpha_2^2 > 0, \quad d = I_3(I_2 - I_1) > 0,$$

and I_i the principal moments of inertia. Hence, the transformation is

$$L = \sqrt{I_3}\sqrt{\frac{a + b\sin^2 \ell}{c + d\sin^2 \ell}}, \quad \beta_1 = -t + \frac{\partial \mathcal{U}}{\partial \alpha_1},$$

$$G = \frac{\partial \mathcal{S}}{\partial g} = \alpha_2, \qquad \beta_2 = g + \frac{\partial \mathcal{U}}{\partial \alpha_2},$$

$$H = \frac{\partial \mathcal{S}}{\partial h} = \alpha_3, \qquad \beta_3 = h,$$

and

$$\mathcal{H} = -\frac{\partial \mathcal{S}}{\partial t} = \alpha_1.$$

Once obtained this transformation, Sadov proceeds to obtain a second one

$$(\beta_1, \beta_2, \beta_3, \alpha_1, \alpha_2, \alpha_3) \longrightarrow (\varphi_\ell, \varphi_g, \varphi_h, I_\ell, I_g, I_h), \tag{18}$$

such that the new actions are

$$I_\ell = \frac{1}{2\pi}\oint L\, d\ell, \quad I_g = \frac{1}{2\pi}\oint G\, dg = G = \alpha_2, \quad I_h = \frac{1}{2\pi}\oint H\, dh = H = \alpha_3.$$

To compute these quadratures, Sadov defines a parameter κ and a state function λ as

$$\kappa^2 = \frac{I_3(I_2 - I_1)}{I_1(I_3 - I_2)} \geq 0, \qquad \lambda^2 = \kappa^2 \frac{I_1}{I_3} \frac{2\, I_3\alpha_1 - \alpha_2^2}{\alpha_2^2 - 2\, I_1\alpha_1} \geq 0. \tag{19}$$

With these quantities, the angular moment L may be put as

$$L^2 = \alpha_2^2 \frac{\kappa^2}{\kappa^2 + \lambda^2} \frac{1 - \lambda^2 + (\kappa^2 + \lambda^2)\sin^2 \ell}{1 + \kappa^2 \sin^2 \ell} = I_g^2\, L_1^2(\ell, \lambda), \tag{20}$$

and by means of the change

$$\sin \ell = \frac{\cos z}{\sqrt{1 + \kappa^2 \sin^2 z}}, \qquad \cos \ell = -\frac{\sqrt{1 + \kappa^2}\sin z}{\sqrt{1 + \kappa^2 \sin^2 z}}, \tag{21}$$

and after some algebra, there results

$$I_\ell = \frac{1}{2\pi}\oint L\, d\ell = \frac{2\alpha_2\sqrt{1 + \kappa^2}}{\pi\kappa\sqrt{\kappa^2 + \lambda^2}} \Big((\kappa^2 + \lambda^2)\Pi(\kappa^2, \lambda) - \lambda^2 K(\lambda)\Big). \tag{22}$$

where K and Π are the complete elliptical integrals of the first and the third kind.

Since the Hamiltonian is $\mathcal{H} = \alpha_1$, we have that

$$\mathcal{H} = \frac{I_g^2}{2\, I_1 I_3} \frac{I_3\lambda^2 + I_1\kappa^2}{\kappa^2 + \lambda^2}. \tag{23}$$

Let $\mathcal{W}$ be the generating function of the transformation from the Serret-Andoyer variables to the action and angle variables. It may be chosen as

$$\mathcal{W} = I_g g + I_h h + \mathcal{V}(\ell; I_\ell, I_g),$$

and the equations of the transformation are

$$\begin{aligned} L &= \frac{\partial \mathcal{V}}{\partial \ell}, & \varphi_\ell &= \frac{\partial \mathcal{V}}{\partial I_\ell}, \\ G &= \frac{\partial \mathcal{W}}{\partial g} = I_g, & \varphi_g &= \frac{\partial \mathcal{W}}{\partial I_g} = g + \frac{\partial \mathcal{V}}{\partial I_g}, \\ H &= \frac{\partial \mathcal{W}}{\partial h} = I_h, & \varphi_g &= \frac{\partial \mathcal{W}}{\partial I_h} = h. \end{aligned}$$

The generating function $\mathcal{V}$ is obtained directly from the quadrature

$$\mathcal{V}(\ell; I_\ell, I_g) = \int_{\ell_0}^{\ell} L(\ell; I_\ell, I_g)\, d\ell,$$

where we have to replace L by its value from expression (20).

To compute φ_ℓ and φ_g , we need before the value of the partial derivatives $\partial\lambda/\partial I_\ell$ and $\partial\lambda/\partial I_g$, since by the rule chain, and taking into account (20), we get

$$\begin{aligned} \varphi_\ell &= \frac{\partial \mathcal{V}}{\partial I_\ell} = \frac{\partial \mathcal{V}}{\partial \lambda}\frac{\partial \lambda}{\partial I_\ell} = \frac{\partial \lambda}{\partial I_\ell}\int_{\ell_0}^{\ell} \frac{\partial L}{\partial \lambda}\, d\ell, \\ \varphi_g &= g + \frac{\partial}{\partial I_g}\int_{\ell_0}^{\ell} L\, d\ell = g + \frac{1}{I_g}\int_{\ell_0}^{\ell} L\, d\ell + \frac{\partial \lambda}{\partial I_g}\int_{\ell_0}^{\ell} \frac{\partial L}{\partial \lambda}\, d\ell, \end{aligned} \tag{24}$$

With the help of the change (21), we compute the quadratures

$$\int_{\ell_0}^{\ell} \frac{\partial L}{\partial \lambda}\, d\ell = -\frac{I_g \kappa \lambda (1+\kappa^2)^{1/2}}{(\kappa^2+\lambda^2)^{3/2}} F(z, \lambda),$$

and

$$\int_{\ell_0}^{\ell} L\, d\ell = \frac{I_g}{\kappa}\sqrt{\frac{1+\kappa^2}{\kappa^2+\lambda^2}}\Big[(\kappa^2+\lambda^2)\Pi(z, \kappa^2, \lambda) - \lambda^2 K(\lambda)\Big].$$

For the other partial derivatives, we put the expression (22) as

$$I_\ell = I_g f(\lambda),$$

hence,

$$\frac{\partial I_\ell}{\partial \lambda} = I_g \frac{\partial f}{\partial \lambda} = \frac{1}{2\pi}\oint \frac{\partial L}{\partial \lambda} = -\frac{I_g 2\kappa\lambda(1+\kappa^2)^{1/2}}{\pi(\kappa^2+\lambda^2)^{3/2}} K(\lambda)$$

and

$$\frac{\partial I_g}{\partial \lambda} = -\frac{I_g}{f(\lambda)}\frac{\partial f}{\partial \lambda} = -\frac{I_g^2}{I_\ell}\frac{\partial f}{\partial \lambda} = \frac{I_g \kappa^2 \lambda K(\lambda)}{(\kappa^2+\lambda^2)[(\kappa^2+\lambda^2)\Pi(\kappa^2, \lambda) - \lambda^2 K(\lambda)]}.$$

Replacing all these expressions into (24), one gets

$$\begin{aligned} \varphi_\ell &= \frac{\pi}{2}\frac{F(z,\lambda)}{K(\lambda)}, \\ \varphi_g &= g + \frac{1}{\kappa}\sqrt{(\kappa^2+\lambda^2)(1+\kappa^2)}\left[\Pi(z,\kappa^2,\lambda) - \frac{\Pi(\kappa^2,\lambda)F(z,\lambda)}{K(\lambda)}\right]. \end{aligned} \tag{25}$$

By inverting the first of these expressions we get

$$z = \mathrm{am}(2K(\lambda)/\pi\varphi_\ell,\, \lambda)$$

and thus, we have the Serret-Andoyer angle g in terms of the action-angle variables

$$g = \varphi_g + \frac{1}{\kappa}\sqrt{(\kappa^2+\lambda^2)(1+\kappa^2)}\left[\frac{2}{\pi}\Pi(\kappa^2,\lambda)\,\varphi_\ell - \Pi\Big(\mathrm{am}(2K(\lambda)/\pi\varphi_\ell,\,\lambda),\kappa^2,\lambda\Big)\right],$$

and the angle ℓ is obtained directly from the expression (21).

3.4 Deprit-Elipe variables

An approach to the complete reduction of the Euler-Poinsot problem, different from the solutions of Sadov or Kinoshita, was given by Deprit and Elipe [5] by a conservative canonical transformation, which is a variant of the Hamilton-Jacobi equation, the so called Delaunay transformation.

Essentially, it consists of a transformation

$$(\delta,\gamma,\Delta,\Gamma) \longrightarrow (\ell,g,L,G)$$

that converts the Hamiltonian (17) into a function $\Phi(\Delta,\Gamma)$ depending only on the momenta. The function chosen in [5] is

$$\Phi = \frac{1}{2}\frac{\Gamma^2}{\Delta}, \tag{26}$$

thus, the dimensions of Γ and Δ are those of the angular momentum and a moment of inertia, respectively. Then, the Delaunay equation reads

$$(a_1\sin^2\ell + a_2\cos^2\ell)\left(\frac{\partial\mathcal{S}}{\partial g}\right)^2 + (a_3 - a_1\sin^2\ell - a_2\cos^2\ell)\left(\frac{\partial\mathcal{S}}{\partial \ell}\right)^2 = \frac{\Gamma^2}{\Delta}.$$

Since the angle g is ignorable, the generating function $\mathcal{S}$ may be put as

$$\mathcal{S} = g\Gamma + \mathcal{W}(\ell;\Delta,\Gamma),$$

and the equations of the transformation are

$$\begin{aligned} \delta &= \frac{\partial\mathcal{S}}{\partial\Delta}, & G &= \frac{\partial\mathcal{S}}{\partial g} = \Gamma, \\ \gamma &= \frac{\partial\mathcal{S}}{\partial\Gamma}, & L &= \frac{\partial\mathcal{S}}{\partial \ell}. \end{aligned}$$

For the region of circulation around the axis of maximal inertia ($a_3 < \mathcal{A} < a_2 < a_3$), there results that the generating function $\mathcal{W}$ is obtained by a quadrature from the relation

$$\left(\frac{\partial \mathcal{W}}{\partial \ell}\right)^2 = \frac{(a_1 \sin^2 \ell + a_2 \cos^2 \ell - \mathcal{A})}{(a_3 - a_1 \sin^2 \ell - a_2 \cos^2 \ell)} \Gamma^2.$$

Proceeding in a similar way as in Section 2, there results that

$$\begin{aligned}
\delta &= \frac{\Gamma}{2\Delta^2} \frac{1}{\sqrt{(a_1 - \mathcal{A})(a_2 - a_3)}} F(\psi, k_3), \\
\gamma &= g + \frac{a_2 - \mathcal{A}}{\sqrt{(a_1 - \mathcal{A})(a_2 - a_3)}} \Pi(\phi, \alpha_3, k_3)
\end{aligned}$$

for the coordinates, and

$$G = \Gamma, \qquad L = \Gamma \sqrt{\frac{a_1 - \mathcal{A}}{a_1 - a_3}} \sqrt{\frac{1 - \alpha_3 \cos^2 \ell}{1 - \alpha_3^0 \cos^2 \ell}}$$

for the moments.

The parameters α_3^0, α_3 and k_3 are

$$\alpha_3^0 = \frac{a_1 - a_2}{a_1 - a_3}, \qquad \alpha_3 = \frac{a_1 - a_2}{a_1 - \mathcal{A}}, \qquad k_3^2 = \frac{\alpha_3 - \alpha_3^0}{1 - \alpha_3^0},$$

and the angles ϕ and ψ are given by the relations

$$\begin{aligned}
\sin \phi &= \frac{\sin \ell}{\sqrt{1 - \alpha_3 \cos^2 \ell}}, & \cos \phi &= \frac{\sqrt{1 - \alpha_3} \cos \ell}{\sqrt{1 - \alpha_3 \cos^2 \ell}}, \\
\cos \psi &= \frac{\sin \ell}{\sqrt{1 - \alpha_3^0 \cos^2 \ell}}, & \sin \psi &= \frac{\sqrt{1 - \alpha_3^0} \cos \ell}{\sqrt{1 - \alpha_3^0 \cos^2 \ell}}.
\end{aligned}$$

4 Quaternions

4.1 Rotations through quaternions

As we saw, a rotation about an axis $\boldsymbol{a}$ and angle ϕ is an element of the group $SO(3)$, and may be expressed in terms of the 3-1-3 sequence of Euler angles. However, this representation is singular for the nutation angle $\vartheta = 0$. This singularity is essential, in the sense that if we change of basis, we meet another singularity. One way of avoiding such singularities is to use quaternions. By using this set, there are no singularities, but we pay a penalty, for the quaternions set is a four dimensional space, thus, we have an extra variable. But we have also an important advantage, namely that by using quaternions we deal only with elementary arithmetic operations, namely addition and multiplication, which speed up the computation.

We may define the quaternion set as

$$\mathcal{Q} = \{q = (q_0; \boldsymbol{q}) \,|\, q_0 \in \mathbb{R},\ \boldsymbol{q} = (q_1, q_2, q_3) \in \mathbb{R}^3\},$$

and in it, we define the following operations

- Addition: $p + q = (p_0 + q_0; \boldsymbol{p} + \boldsymbol{q})$, with $p, q \in \mathcal{Q}$.
- Multiplication by an scalar: $\lambda q = \lambda q_0 + \lambda \boldsymbol{q}$, with $q \in \mathcal{Q}$ and $\lambda \in \mathbb{R}$.
- Multiplication: $pq = (p_0 q_0 - \boldsymbol{p} \cdot \boldsymbol{q}; q_0 \boldsymbol{p} + p_0 \boldsymbol{q} + \boldsymbol{p} \times \boldsymbol{q})$ with $p, q \in \mathcal{Q}$,

that define an structure of algebra on the quaternion set.

The conjugate quaternion $\tilde{q}$ of a quaternion q is defined as $\tilde{q} = (q_0; \boldsymbol{q})$. With this, the norm of a quaternion q is

$$\|q\| = \sqrt{q\tilde{q}} = (q_0^2 + q_1^2 + q_2^2 + q_3^2)^{1/2},$$

and the quaternions which norm is one form the subalgebra of unitary quaternions.

Those vectors with real part, that is, $q_0 = 0$, will be denoted as *pure vectors*.

Let us consider the rotation $R(\phi, \boldsymbol{a})$ of an angle ϕ about an axis $\boldsymbol{a}$, with $\|\boldsymbol{a}\| = 1$ (the components of this vector are commonly known as direction cosines). We can build the unit quaternion

$$q = (\cos(\phi/2); \sin(\phi/2)\, \boldsymbol{a}). \tag{27}$$

Thus, if we have a unit quaternion, this formula provide us the elements of the rotation, the axis and the argument.

Let us now assume that we apply the above rotation to a vector $\boldsymbol{x}$. We define the quaternion $x = (0; \boldsymbol{x})$, and by means of the definition of the product of quaternions, we easily find that

$$q\, x\, \tilde{q} = (0; \boldsymbol{x} + 2q_0(\boldsymbol{q} \times \boldsymbol{x}) + 2\boldsymbol{q} \times (\boldsymbol{q} \times \boldsymbol{x})),$$

but its vectorial part is precisely the well known Euler's formula for the image of $\boldsymbol{x}$ by the rotation $R(\phi, \boldsymbol{a})$, that is,

$$R(\phi, \boldsymbol{a})\boldsymbol{x} = \boldsymbol{x} + (\boldsymbol{a} \times \boldsymbol{x}) \sin\phi + \boldsymbol{a} \times (\boldsymbol{a} \times \boldsymbol{x})(1 - \cos\phi).$$

Thus, the image of a vector $\boldsymbol{x}$ by a rotation $R(\phi, \boldsymbol{a})$ is obtained by the product of quaternions $q\, x\, \tilde{q}$,

$$R(\phi, \boldsymbol{a})\boldsymbol{x} = q\, x\, \tilde{q} \tag{28}$$

It is easy to check that the composition of two rotations $R(\psi, \boldsymbol{b}) \circ R(\phi, \boldsymbol{a})$ is given by the quaternion product pq of the corresponding to each rotation p and q. Hence by using quaternions, there is no need to evaluate trigonometric functions; the only operations needed are the elemental arithmetic ones.

We can obtain the matrix M of the rotation by building up a matrix which columns are the images of the vectors of the basis by the rotation, that is, the i-column will be the components of the vector part of the quaternion $q\boldsymbol{e}_i\tilde{q}$. Thus, we have

	$\boldsymbol{e}_1$	$\boldsymbol{e}_2$	$\boldsymbol{e}_3$
$R(\boldsymbol{e}_1)$	$q_1^2 - q_2^2 - q_3^2 + q_0^2$	$2(q_1q_2 + q_0q_3)$	$2(q_1q_3 - q_0q_2)$
$R(\boldsymbol{e}_2)$	$2(q_1q_2 - q_0q_3)$	$q_2^2 - q_3^2 - q_1^2 + q_0^2$	$2(q_2q_3 + q_0q_2)$
$R(\boldsymbol{e}_3)$	$2(q_1q_3 + q_0q_2)$	$2(q_2q_3 - q_0q_1)$	$q_3^2 - q_1^2 - q_2^2 + q_0^2$

Conversely, from this expression, we can extract the quaternion of a rotation from the matrix $(r_{i,j})$. Indeed, one deduces immediately that

$$\begin{aligned} 4q_0^2 &= r_{1,1} + r_{2,2} + r_{3,3} + 1, \\ 4q_1^2 &= r_{1,1} - r_{2,2} - r_{3,3} + 1, \\ 4q_2^2 &= r_{2,2} - r_{3,3} - r_{1,1} + 1, \\ 4q_3^2 &= r_{3,3} - r_{1,1} - r_{2,2} + 1. \end{aligned}$$

Albeit these formulas are correct, they are inadequate for components that happen to be small, since in taking the square root of a small number we can loss significance. To avoid it, we use the preceding formulas to compute the biggest component of the quaternion, and the other three components are computed, accordingly, from three of the following the relations

$$\begin{aligned} r_{1,2} + r_{2,1} &= 4q_1q_2, & r_{1,2} - r_{2,1} &= 4q_3q_0, \\ r_{2,3} + r_{3,2} &= 4q_2q_3, & r_{2,3} - r_{3,2} &= 4q_1q_0, \\ r_{3,1} + r_{1,3} &= 4q_3q_1, & r_{3,1} - r_{1,3} &= 4q_2q_0. \end{aligned}$$

Several methods for finding the quaternion from the rotation matrix have been developed recently. See e.g. [16].

4.2 Canonical extension to quaternions

Because of the above mentioned properties of the quaternions, they are used to formulate the dynamics of the rigid body. Euler equations do not present much difficulties, however, the Hamilton equations are quite more involved since quaternions represent an extension of the dimensions of the Hamiltonian system. For that, several solutions have been proposed, for instance, the ones given in [17-20].

Lowering the dimensions of a Hamiltonian system is a step most frequent in Mechanics. Given a Hamiltonian $\mathcal{H}(\boldsymbol{p}, \boldsymbol{P})$ in n coordinates $\boldsymbol{p}$ and n momenta $\boldsymbol{P}$, one build a canonical transformation

$$\lambda : (\boldsymbol{q}, \boldsymbol{Q}) \longrightarrow (\boldsymbol{p}, \boldsymbol{P}) \; : \mathbb{R}^n \times \mathbb{R}^n \longrightarrow \mathbb{R}^n \times \mathbb{R}^n$$

that will render $m < n$ coordinates ignorable, $(q_i)_{1\leq i\leq m}$ in the pullback of the Hamiltonian $\mathcal{H}$. Thus, there are m integrals, namely the conjugate momenta $(Q_i)_{1\leq i\leq m}$, and the problem is reduced to another one which dimension is $2m$ lower.

The reverse problem, that of increasing the dimension of a Hamiltonian system appear in orbital dynamics (21) with the K-S transformation. Assume that a transformation λ maps a domain of dimension $2(n+m)$ in the phase space $(\boldsymbol{q}, \boldsymbol{Q})$ into a domain of dimension $2n$ in the phase space $(\boldsymbol{p}, \boldsymbol{P})$. Provided that the Poisson brackets

$$\{p_i; p_j\} = \{P_i; P_j\} = 0 \quad \text{for} \quad 1 \leq i \leq j \leq n,$$

and that exists a so-called multiplier μ independent of $\boldsymbol{q}$ and $\boldsymbol{Q}$ such that

$$\{p_i; P_j\} = \mu\delta_{i,j} \quad \text{for} \quad 1 \leq i \leq j \leq n,$$

with $\delta_{i,j}$ the Kronecker symbol, then

$$\lambda^* \boldsymbol{p}(t) = \boldsymbol{p}(\boldsymbol{q}(t), \boldsymbol{Q}(t)) \quad \text{and} \quad \lambda^* \boldsymbol{P}(t) = \boldsymbol{P}(\boldsymbol{q}(t), \boldsymbol{Q}(t))$$

is solution of the Hamilton equations

$$\dot{\boldsymbol{p}} = \boldsymbol{\nabla}_{\boldsymbol{P}}\mathcal{H}, \qquad \dot{\boldsymbol{P}} = -\boldsymbol{\nabla}_{\boldsymbol{p}}\mathcal{H},$$

whenever $(\boldsymbol{q}(t), \boldsymbol{Q}(t))$ is a solution of the system

$$\dot{\boldsymbol{q}} = \boldsymbol{\nabla}_{\boldsymbol{Q}}(\lambda^*(\mu^{-1}\mathcal{H})), \qquad \dot{\boldsymbol{Q}} = -\boldsymbol{\nabla}_{\boldsymbol{q}}(\lambda^*(\mu^{-1}\mathcal{H})).$$

Such a transformation is called *weakly* canonical. (For an extension of this theorem, see Appendix in [22]). This result gives us a way to check whether a transformation is weakly canonical or not, but does not supply a technique for building canonical transformations capable of raising the dimension of the Hamiltonian system. This is complemented in [18], where it is stated that a transformation $\boldsymbol{q} = \boldsymbol{q}(\boldsymbol{p})$ given raising by one the dimension, its extension with the moments obtained by

$$Q_i = \left(\frac{\partial \boldsymbol{p}}{\partial q_i}\right) \cdot \boldsymbol{P}, \qquad (1 \leq i \leq n+1), \tag{29}$$

is weakly canonical.

Coming back to the Euler sequence 3-1-3, one defines the so-called Euler parameters by the relations

$$\begin{aligned} q_0 &= \cos\frac{\vartheta}{2}\cos\frac{\phi+\psi}{2}, & q_1 &= \sin\frac{\vartheta}{2}\cos\frac{\phi-\psi}{2}, \\ q_2 &= \sin\frac{\vartheta}{2}\sin\frac{\phi-\psi}{2}, & q_3 &= \cos\frac{\vartheta}{2}\sin\frac{\phi+\psi}{2}. \end{aligned} \tag{30}$$

After some algebra, one finds that inverse transformation is

$$\tan\phi = \frac{q_1q_3 + q_0q_2}{q_0q_1 - q_2q_3}, \qquad \tan\varphi = \frac{\sqrt{q_1^2 + q_2^2}}{\sqrt{q_0^2 + q_3^2}}, \qquad \tan\psi = \frac{q_1q_3 - q_0q_2}{q_0q_1 + q_2q_3}. \tag{31}$$

Then, we can use these relations to compute the partial derivatives of the Euler angles with respect to the Euler parameters and thus to obtain the conjugate moments to the Euler parameters by the relations (29) as

$$\begin{aligned}
Q_0 &= -(\Phi + \Psi)\cos\frac{\vartheta}{2}\sin\frac{\phi+\psi}{2} - 2\Theta\sin\frac{\vartheta}{2}\cos\frac{\phi+\psi}{2}, \\
Q_1 &= -(\Phi + \Psi)\sin\frac{\vartheta}{2}\sin\frac{\phi-\psi}{2} + 2\Theta\cos\frac{\vartheta}{2}\cos\frac{\phi-\psi}{2}, \\
Q_2 &= (\Phi + \Psi)\sin\frac{\vartheta}{2}\cos\frac{\phi-\psi}{2} + 2\Theta\cos\frac{\vartheta}{2}\sin\frac{\phi-\psi}{2}, \\
Q_3 &= (\Phi + \Psi)\cos\frac{\vartheta}{2}\cos\frac{\phi+\psi}{2} - 2\Theta\sin\frac{\vartheta}{2}\sin\frac{\phi+\psi}{2}.
\end{aligned} \tag{32}$$

With the help of a symbolic processor, one easily checks that the Poisson brackets satisfy the conditions for the transformation be weakly canonical.

From these expressions, one readily finds that

$$\begin{aligned}
\Phi &= \frac{1}{2}(-q_3Q_0 - q_2Q_1 + q_1Q_2 + q_0Q_3), \\
\Theta &= \frac{1}{2\sqrt{q_1^2 + q_2^2}\sqrt{q_0^2 + q_3^2}}(-q_0Q_0 + q_1Q_1 + q_2Q_2 - q_3Q_3), \\
\Psi &= \frac{1}{2}(-q_3Q_0 + q_1Q_1 + q_2Q_2 - q_3Q_3).
\end{aligned} \tag{33}$$

By replacing Eqs. (31) and (34) into the expressions of the components of the angular momentum (5), and after some algebraic simplifications, there results that

$$\begin{aligned}
g_1 &= \frac{1}{2}(-q_1Q_0 + q_0Q_1 + q_3Q_2 - q_2Q_3), \\
g_2 &= \frac{1}{2}(-q_2Q_0 - q_3Q_1 + q_0Q_2 + q_1Q_3), \\
g_3 &= \frac{1}{2}(-q_3Q_0 + q_2Q_1 - q_1Q_2 + q_0Q_3).
\end{aligned}$$

On the other hand, one easily checks that

$$q_0^2 + q_1^2 + q_2^2 + q_3^2 = 1, \tag{34}$$

and also that

$$q_0Q_0 + q_1Q_1 + q_2Q_2 + q_3Q_3 = 0.$$

In other words, Euler parameters are on the $\mathcal{S}^3$ sphere, and their conjugate moments belong to the tangent plane to the sphere.

The above relations suggest to use the algebra of quaternions. Thus, we can define the quaternion $q = (q_0; \boldsymbol{q}) = (q_0; (q_1, q_2, q_3))$, that obviously is a unit quaternion. Besides, we define the quaternion $Q = (Q_0; \boldsymbol{Q}) = (Q_0; (Q_1, Q_2, Q_3))$ (note that the latter is not unitary). By means of these quaternions, there results that the product $\tilde{q}Q$ is a pure vector, and its vector part is precisely twice the angular moment in the body frame, that is,

$$\frac{1}{2}\tilde{q}Q = (0; \boldsymbol{g}) = (0; \boldsymbol{b}_1 g_1 + \boldsymbol{b}_2 g_2 + \boldsymbol{b}_3 g_3).$$

It is worth noting that the angular moment in the space frame is given by

$$\frac{1}{2}Q\tilde{q} = (0; \boldsymbol{G}) = (0; \boldsymbol{s}_1 G_1 + \boldsymbol{s}_2 G_2 + \boldsymbol{s}_3 G_3).$$

As one could expect, a rotation may be represented by a quaternion (28),

$$q(\frac{1}{2}\tilde{q}Q)\tilde{q} = \frac{1}{2}Q\tilde{q}.$$

Let us build now the quaternions P and Q as the pure vectors

$$P = (0; (\sqrt{a_1}g_1, \sqrt{a_1}g_2, \sqrt{a_1}g_3)) \qquad \text{and} \qquad S = (0; (a_1 g_1, a_1 g_2, a_1 g_3)),$$

then, the Hamiltonian (7) of the rigid body in free rotation is

$$\mathcal{H} = \frac{1}{2}P\tilde{P}, \tag{35}$$

and the Hamilton equations take the simple and compact form

$$\dot{q} = \frac{1}{2}qS, \qquad \dot{Q} = \frac{1}{2}QS.$$

4.3 Quaternions and Serret-Andoyer variables

As it happened with Euler angles, that is not a regular representation for the group $SO(3)$, the same happens for the Serret-Andoyer variables. Thus, we can use the quaternions to remedy this deficiency [5]. For the Serret-Andoyer, let us define the quaternion $p = (p_0;\ (p_1, p_2, p_3))$ defined by

$$\begin{aligned} p_0 &= \sqrt{2(G+L)}\cos\frac{g+\ell}{2}, \qquad & p_2 &= \sqrt{2(G-L)}\sin\frac{g-\ell}{2}, \\ p_1 &= \sqrt{2(G-L)}\cos\frac{g-\ell}{2}, \qquad & p_3 &= \sqrt{2(G-+L)}\sin\frac{g+\ell}{2}. \end{aligned} \tag{36}$$

It is a matter of computing Poisson brackets to show that

$$\begin{aligned} &\{p_3; p_0\} = \{p_2; p_1\} = 1, \\ &\{p_3; p_1\} = \{p_3; p_2\} = \{p_2; p_0\} = \{p_1; p_0\} = 0, \end{aligned}$$

thus, the transformation

$$(\ell, g, L, G) \longrightarrow (p_3, p_2, p_0, p_1)$$

is canonical; the variables p_3 and p_2 are the coordinates, and p_0 and p_1 are their respective conjugate moments. Note that this quaternion is not unitary; its norm is $\|p\| = 2\sqrt{G}$.

After some algebra, one gets that the angular momentum vector $\boldsymbol{G}$ is the vectorial part of the quaternion $[\tilde{p}\,(0; (0,0,1)]\,p)/4$, that is,

$$\begin{aligned} g_1 &= (-p_0 p_2 + p_1 p_3)/2, \\ g_2 &= (p_0 p_1 + p_1 p_3)/2, \\ g_3 &= (p_0^2 - p_1^2 - p_2^2 + p_3^2)/4. \end{aligned}$$

Thus, by defining a quaternion $W = (0\,;\,\boldsymbol{G})$, one readily checks that the equations of Hamilton for the set of canonical variables defined by the quaternion p take the simple and compact form

$$\dot{p} = \frac{1}{2} p\,\tilde{W}.$$

5 Motion under a Newtonian Gravity Field

Let us consider a rigid body S, with center of mass C and mass m, that is moving under the gravitational influence of a point O of mass M. Besides the space $\mathcal{S}(C, \boldsymbol{s}_1, \boldsymbol{s}_2, \boldsymbol{s}_3)$ and the body $\mathcal{B}(C, \boldsymbol{b}_1, \boldsymbol{b}_2, \boldsymbol{b}_3)$ reference systems defined in Section 2, let us consider another inertial frame $\mathcal{S}'(O, \boldsymbol{s}_1, \boldsymbol{s}_2, \boldsymbol{s}_3)$, parallel to the space system and with origin at the point O. As it is well known from Chasles (see e.g. [23, p. 2-4]), *"the most general displacement of a rigid body can be obtained by first translating the body, and then rotating it about a line."*

Under the assumption that the dimensions of the rigid body are small enough with respect to the dimensions of the relative orbit of C around O, and that the mass of S is small with respect to the mass of O, it is possible to assume that the orbital motion is independent of the rotation of S [24]. Besides, under these hypotheses, the gravitational potential U may be considered as a perturbation of the kinetic energy of rotation T. Hence, our problem will consist of finding the attitude of the rigid body, assuming that the orbital motion is a known function of the time.

When the rigid body has axial symmetry, $a_1 = a_2$, Serret-Andoyer variables are very convenient, since the angular variable ℓ does not appear in (17) and their Hamilton's equations are immediately integrated. In cases where the axial symmetry "almost" exists, i.e., when $a_1 \approx a_2$, some authors [25], by means of trigonometric relations, split the Hamiltonian (17) in two parts, one independent of the angle ℓ, and a second one depending on ℓ, but that have the small quantity $(a_1 - a_2)$ as factor, and that is considered as a perturbation. In this way, elliptic functions are avoided.

However, in general, the body is far from the axial symmetry, such as it happens in most of asteroids, and the previous assumption ($a_1 \approx a_2$) is no longer valid. Thus, it is necessary to cope with elliptic functions.

5.1 Expansion of the potential function

In order to apply a general perturbation method, it is desirable to have the unperturbed Hamiltonian (17) the simplest possible in order the Lie derivative $\mathcal{L}_0$, or the Poisson bracket associated to the unperturbed Hamiltonian $\mathcal{H}_0$, be easy to evaluate. In this section, we shall use Sadov's variables, although Kinoshita's of the ones given by Deprit and Elipe will serve to our purposes.

The potential between the point O and the rigid body S may be represented up to the second order by means of the well known MacCullagh's formula (see e.g. [26]):

$$V = -\mathcal{G}\,\frac{mM}{r} - \mathcal{G}\,\frac{M}{2r^3}\left[(I_1 - I_2)(1 - 3\alpha^2) + (I_3 - I_2)(1 - 3\gamma^2)\right], \qquad (37)$$

where $\mathcal{G}$ is the gravitational constant, r is the distance between the centers of mass O and C, and (α, β, γ) the direction cosines of the unit position vector $\boldsymbol{u} = \boldsymbol{x}/r$ expressed in the body frame $\mathcal{B}$, that is to say,

$$\boldsymbol{u} = \boldsymbol{x}/r = \boldsymbol{b}_1\alpha + \boldsymbol{b}_2\beta + +\boldsymbol{b}_3\gamma.$$

The presence of the direction cosines suggests to choose an orbital moving frame $\mathcal{O}(\boldsymbol{u}, \boldsymbol{v}, \boldsymbol{w})$, where $\boldsymbol{w}$ is in the direction of the orbital angular moment ($\boldsymbol{\Theta} = \boldsymbol{x} \times \dot{\boldsymbol{x}}$), and $\boldsymbol{v} = \boldsymbol{w} \times \boldsymbol{u}$. In this way, the two bases $\mathcal{O}$ and $\mathcal{B}$ are related by several compositions of rotations involving as arguments both sets of angular variables the Serret-Andoyer (ℓ, g, h) and Polar-nodal (r, θ, ν) (see Figure 3). In matrix form, there results

$$\begin{pmatrix} \alpha \\ \beta \\ \gamma \end{pmatrix} = R(-\ell, \boldsymbol{b}_3)\circ R(-\sigma, \boldsymbol{m})\circ R(-g, \boldsymbol{n})\circ R(-\epsilon, \boldsymbol{\ell})\circ R(\nu - h, \boldsymbol{s}_3)\circ R(I, \boldsymbol{N})\circ R(\theta, \boldsymbol{w}) \begin{pmatrix} 1 \\ 0 \\ 0 \end{pmatrix},$$

where $\boldsymbol{N}$ and I are the node direction and inclination of the orbital plane $(\boldsymbol{uv})$ with respect to the space plane $(\boldsymbol{s}_1\boldsymbol{s}_2)$.

Once obtained α and γ by means of the previous composition of matrices, it is necessary to square their expressions and replace them in equation (37). Thus, in the potential function will appear expressions of the type

$$\prod_{a,b} \sin^a \cos^b,$$

that is, combinations of products of powers of trigonometric sines and cosines of the involved variables.

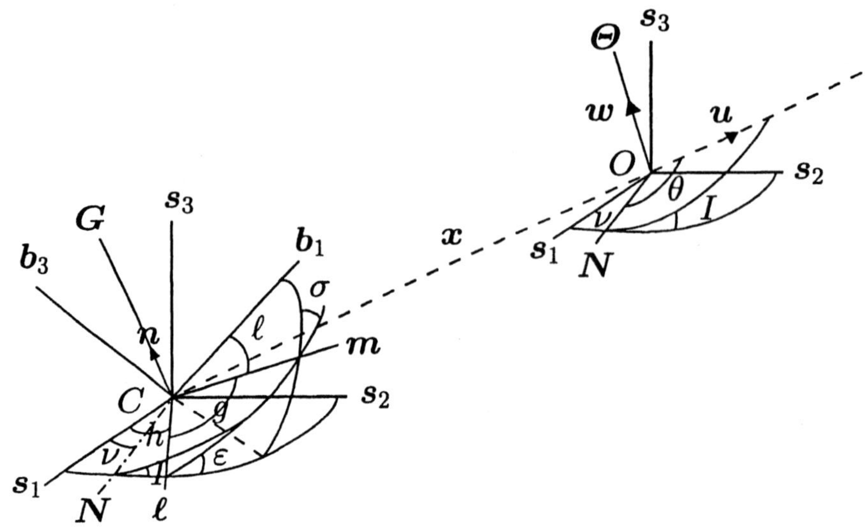

Figure 3: Orbital–rotational variables.

The computation of the potential (37) has been previously carried out by several authors (e.g. [4,27]), although they avoid the powers in favor of linear combinations of angles. Here we maintain the powers.

Note that the potential is given in terms of the Serret-Andoyer variables, whereas the kinetic energy was expressed in terms of the action and angle variables, namely, Sadov's variables. Thus, it is necessary to substitute the solution obtained for the free rotation, that is given in terms of elliptic functions. Hence, we deal with expressions of the type

$$\prod_{\alpha,\beta} \mathrm{sn}^a(z,k)\,\mathrm{cn}^b(z,k)\,\mathrm{dn}^c(z,k). \tag{38}$$

The reason why we maintain the powers of trigonometric functions in the potential, is because we will expand terms of the type appearing in Eq. [38] as Fourier powers series, which coefficients are functions of the Jacobian nome q, that, we recall, is defined by

$$q = q(k) = \exp\left(-\pi K'(k)/K(k)\right).$$

This type of expansions has proved to be very fast convergent, and it is possible to generate them automatically by computer [7] by making use of the properties of the Theta functions. For instance, with the software developed by the authors [7], there results

$$\mathrm{sn}(u,k)\,\mathrm{dn}(u,k) = \frac{\pi^2}{k^2K^2}\sum_{m=0}^{\infty}\frac{(m+1)\,q^{m+1/2}}{1+q^{m+1}}\sin(2m+1)\frac{\pi u}{2K}.$$

The complete expansion of the potential (37) is long enough to be reproduced here, but it can be stored in the hard disk of our computer.

5.2 Analytical solution

So far, for the attitude dynamics of a triaxial rigid body under a Newtonian force field, we obtained the Hamiltonian

$$\mathcal{H} = \mathcal{H}(\varphi_\ell, \varphi_g, -, I_\ell, I_g, I_h) = \mathcal{H}_0 + V = \frac{I_g^2}{2\, I_1 I_3}\,\frac{I_3\lambda^2 + I_1\kappa^2}{\kappa^2 + \lambda^2} + \mathcal{H}_1,$$

in a set of canonical variables $(\varphi_\ell, \varphi_g, \varphi_h, I_\ell, I_g, I_h)$, in which the unperturbed Hamiltonian $\mathcal{H}_0$ depends on the momenta, and the perturbation is a truncated Fourier series, which arguments are the angular variables whereas the coefficients are functions on the momenta. Hence, it is possible to eliminate the variables φ_ℓ, φ_g by means of a Lie-Deprit transformation [28].

Such a transformation, up to the first order,

$$(\varphi_i', I_i') \xrightarrow{\ \mathcal{W}\ } (\varphi_i, I_i) \qquad (i = \ell, g, h),$$

will render the new Hamiltonian $\mathcal{K}$ independent of the coordinates, and hence, will be immediately integrable.

The new Hamiltonian is chosen to be

$$\mathcal{K} = \mathcal{K}(-,-,-, I_\ell', I_g', I_h') = \mathcal{K}_0 + \mathcal{K}_1,$$

with $\mathcal{K}_0 = \mathcal{H}_0$ and $\mathcal{K}_1 = \langle\langle \mathcal{H}_1 \rangle_{\varphi_\ell}\rangle_{\varphi_g}$ the average over the variables φ_ℓ, φ_g. Hence the equations of the motion are

$$\left.\begin{aligned} \dot{I}_i' &= -\frac{\partial \mathcal{K}}{\partial \varphi_i'} = 0, \\ \dot{\varphi}_i' &= \frac{\partial \mathcal{K}}{\partial I_i'} = n_i', \end{aligned}\right\} \Longrightarrow \left.\begin{aligned} I_i' &= \text{constant}, \\ \varphi_i' &= n_i' t + \varphi_i'^{(0)}. \end{aligned}\right\} \tag{39}$$

The generating function $\mathcal{W}_1 = \mathcal{W}_1(\varphi_\ell, \varphi_g, -, I_\ell, I_g, I_h)$ of the transformation must be solution of the PDE

$$\{\mathcal{H}_0; \mathcal{W}_1\} = \mathcal{K}_1 - \mathcal{H}_1. \tag{40}$$

The procedure for both tasks, averaging the Hamiltonian and computing the generating function has been automated by means of an algebraic manipulator. Once obtained the solution for the normalized Hamiltonian (39), it is possible to recover the solution for the original Hamiltonian by means of the equations of the transformation

$$\varphi_i' = \varphi_i - \left.\frac{\partial \mathcal{W}_1}{\partial I_i}\right|_{\varphi_i(0), I_i(0)}, \qquad I_i' = I_i - \left.\frac{\partial \mathcal{W}_1}{\partial \varphi_i}\right|_{\varphi_i(0), I_i(0)}, \qquad (i = \ell, g, h).$$

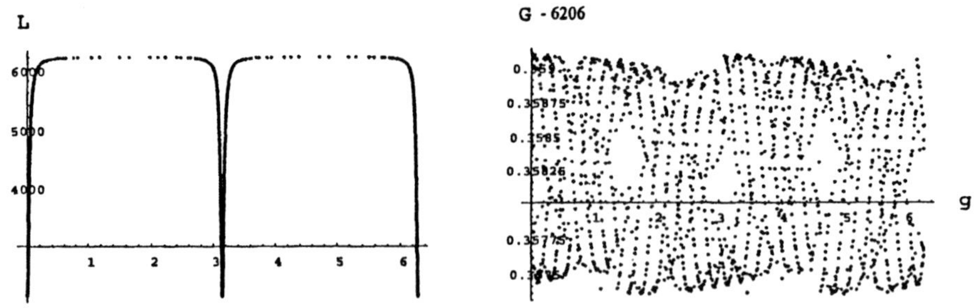

Figure 4: Evolution of the Serret angles ℓ and g versus their respective moments after normalization. Note that normalization is carried out in Sadov's variables, then converted to Serret's ones. Since the graphics represent several revolutions, we observe that ℓ is almost periodic, while g suffers a *precession*-like shift.

5.3 Application to an Eros-like body

As an illustration, we apply the above exposed method to the attitude of a natural body that is far away from the axial symmetry; namely, the asteroid (433) Eros, the target of the NEAR mission [29].

The following orbital elements of (433) Eros are extracted from Minor Planet Circular No. 24086, updated to a current osculation epoch (JDT 2450200.5):

$$a = 1.4583190AU, \quad e = 0.2229864, \quad i = 10^\circ\!.83085,$$
$$\Omega = 304^\circ\!.43847, \quad \omega = 178^\circ\!.55858, \quad P = 1.76 \text{ years}.$$

The closest approach of Eros to Earth in the 20th century was on January 23, 1975, at approximately 0.15 AU. Previous close approaches occurred in 1901 at 0.32 AU and in 1931 at 0.17 AU. Because of its repeated close encounters with Earth, Eros has been an important object historically for refining the mass of the Earth-moon system and the value of the astronomical unit.

Eros also is one of the most elongated asteroids, a "potato-shaped" body with estimated dimensions of 40.5 by 14.5 by 14.1 km. Its size qualifies Eros as one of only three near-Earth asteroids (NEA) with diameters above 10 kilometers. Astronomers assign the asteroid a rotation period of 5.27 hours. The value of the mass adopted for NASA is $5\,10^{15}$ kg.

In addition to these physical characteristics, we choose in rather arbitrary way the inclinations ϵ and σ to have the modulus of the elliptic functions involved in the problem is very high, and close to the unit (the most unfavorable case). With these criteria, in terms of Serret-Andoyer variables, we choose as initial conditions the values

$$\begin{aligned} \ell &= 0.5235987755982988, & L &= 6187.225633746185, \\ g &= 0.5235987755982988, & G &= 6206.357763463754, \\ h &= 0.7853981633974483, & H &= 5374.863488134382, \\ \varepsilon &= 0.5235987755982988, & \sigma &= 0.07853981633974448. \end{aligned}$$

Hence, there results

$$\lambda = 0.936057480558765, \qquad \kappa = 23.77968208793787.$$

Thus, the modulus of the elliptic functions involved in the problem is very high, and close to the unit. In Figure 4 we present the evolution of the Serret-Andoyer variables after normalization that was carried out in Sadov's variables. Since the graphics represent several revolutions, we observe ℓ is almost periodic, while g suffers a *precession*-like shift. The combination of the three components g_1, g_2 and g_3 of the angular momentum vector over the sphere appears in Figure 5. Since g_3 is almost a dn (see [5]), its period is near twice the period of g_1 and g_2 (cn and sn respectively).

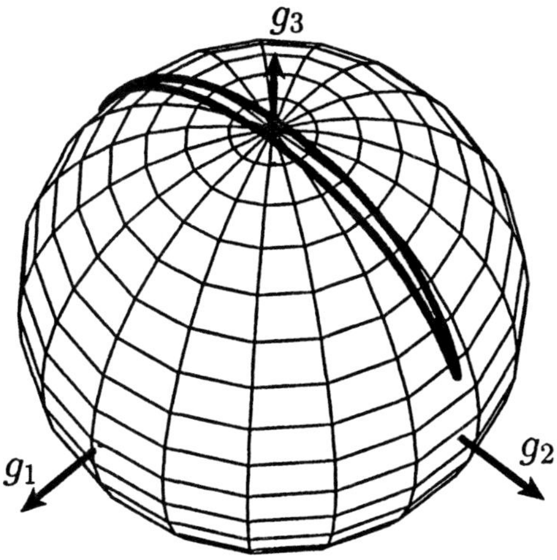

Figure 5: Evolution of the g_i components on the angular momentum sphere.

6 Transversal Homoclinic Orbits

In the previous sections we dealt with rigid bodies. However, in most of the cases, this is only an approximation to the actual cases in which the bodies are not perfect rigid bodies, but they suffer deformations. Besides, the presence of perturbing forces, most often, will break the integrability and will give rise to chaotic behavior [30-32].

In this section, we will show how a perturbation will originate chaos. In particular, that when one moment of inertia is subject to a time periodic perturbation, the system has Smale's horseshoes, the heteroclinic orbits transversally intersects, and the system is chaotic.

Let us assume that one of the moments of inertia of the rigid body is suffering a periodic perturbation. For instance, let us take

$$a_3 = a_3^0 + \epsilon \cos \mu t, \qquad \text{with} \quad \epsilon \ll 1, \quad a_3^0,\, \nu = \text{constant}. \tag{41}$$

Under this hypothesis, the Hamiltonian (7) becomes

$$\mathcal{H} = \mathcal{H}_0 + \mathcal{P} = \frac{1}{2}(a_1 g_1^2 + a_2 g_2^2 + a_3^0 g_3^2) + \frac{1}{2}\epsilon g_3^3 \cos \mu t. \tag{42}$$

It is expected that the perturbation will break the heteroclinic orbits connecting the saddle points of the unperturbed motion, and that transverse heteroclinic orbits appear, which will give rise to chaotic motion. To check that, we use Melnikov's method. Briefly, this method [9] states that if the function

$$M(t_0) = \int_{\infty}^{-\infty} \{\mathcal{H}_0(g_i(t-t_0))\,;\, \mathcal{P}(g_i(t-t_0),\, t)\}\, dt$$

has simple zeroes, the perturbation gives rise to chaotic motion in the sense that the system has Smale's horseshoes.

By means of the symplectic structure (6), there results for the Poisson bracket

$$\{\mathcal{H}_0; \mathcal{P}\} = \epsilon(a_1 - a_2)\, g_1 g_2 g_3 \cos \mu t,$$

and by replacing the solution (14) of the heteroclinic orbit, the Melnikov's function is

$$M(t_0) = \epsilon\, G^3 n_2 \frac{(a_1 - a_2)}{(a_1 - a_3^0)} \int_{-\infty}^{\infty} \frac{\sinh n_2(t-t_0)}{\cosh n_2(t-t_0)} \cos \mu t\, dt.$$

Straightforward calculations and with the help of the classical book with tables of integrals [33], we have

$$M(t_0) = \frac{\epsilon\, G^3 (a_1 - a_2)\pi \mu^2}{2n_2^2 (a_1 - a_3^0) \sinh(\mu\pi/2n_2)} \sin \mu t_0. \tag{43}$$

that have simple zeroes, and therefore, we conclude that the perturbation gives rise to chaotic motion in the sense that the system has Smale's horseshoes. This can be also checked by computing the Poincaré sections for the Hamiltonian (42) for several values of the parameter μ. We see in Figure 6 that stochastic motion appears around the separatrices of the unperturbed motion. For details on the estimation of the stochasticity layer, see [34].

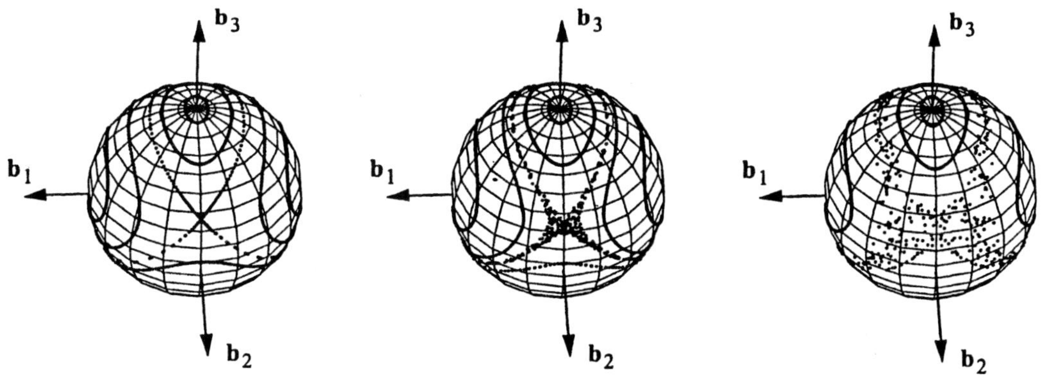

Figure 6: Poincaré surface of section for a tri-axial rigid body with a periodic perturbation of the type $\epsilon g_3^3 \cos \mu t$. The figures are obtained with the values (from left to right) $\mu = 2.0,\ 0.6\ .02$. We see that the heteroclinic orbit is broken and the chaotic motion appears.

7 Gyrostat in Free Motion

As we just saw, the presence of separatrices in the phase flow could create chaotic motion when a perturbation is included. However, there are mechanisms that avoids such non desired motion. This can be achieved by shifting the saddle points to other regions, or by counter-balancing the influence of the perturbation. One of these mechanisms is the *gyrostat*. A gyrostat $\mathcal{G}$, let us recall, is a mechanical system made of a rigid body $\mathcal{P}$ called the *platform*, and other bodies $\mathcal{R}$ called the *rotors*, connected to the platform, and in such a way that the motion of the rotors does not modify the distribution of mass of the gyrostat $\mathcal{G}$. Due to this double spinning, the gyrostat on the one hand and the rotors on the other, the gyrostat is also known with the name of *dual-spin* body.

Besides the classical use of the gyrostats for modelling the rotation of the Earth [35], it is frequently used in Astrodynamics for controlling the attitude dynamics of spacecrafts, and for stabilizing their rotations [36-40].

The simplest case, the gyrostat with constant internal moments in free rotation, is an integrable problem (see e.g. [41]) in terms of elliptic functions. However, the problem is not easy to tackle, for there is a huge number of possibilities, depending on the values of the gyrostatic moments of the rotors and on the values of the principal moments of inertia. Recently, the author [42-46] proved that the dynamical problem of a gyrostat in free rotation is equivalent to a multi parametric quadratic Hamiltonian on the unit sphere. To each gyrostat case (depending on the moments of inertia and the rotors that act simultaneously) corresponds one case of the parametric quadratic Hamiltonian.

Let us denote by g_1, g_2 and g_3 the components of the angular moment $\boldsymbol{G}$ of the gyrostat $\mathcal{G}$ expressed in the body frame. The symplectic structure is the same as in the rigid body, then the Poisson brackets satisfy the relation (6).

Straightforward computations show that the angular momentum, in the principal axes of inertia, is

$$\boldsymbol{G} = \mathbb{I}\,\boldsymbol{\omega} + \boldsymbol{f},$$

where $\mathbb{I}$ is the diagonal inertia tensor, $\boldsymbol{\omega}$ is the angular velocity, and $\boldsymbol{f}$ is the angular momentum of the rotors, that here will be assumed constant.

On the other hand, the kinetic energy is

$$T = \frac{1}{2}\boldsymbol{\omega}\cdot\mathbb{I}\,\boldsymbol{\omega} + \boldsymbol{\omega}\cdot\boldsymbol{f} + T_{\mathcal{R}}, \tag{44}$$

where $T_{\mathcal{R}}$ is the kinetic energy of the rotors in its relative motion, that is supposed to be a function of the time.

Under no external forces, the Hamiltonian, let us remember that is obtained as the Legendre transformation of the kinetic energy, is

$$\mathcal{H} = \frac{1}{2}\boldsymbol{\omega}\cdot\mathbb{I}\,\boldsymbol{\omega}, \tag{45}$$

that does not coincide with the kinetic energy (44), since the latter is not a quadratic form in the velocities. In the present case, the Hamiltonian (45) is precisely the kinetic energy of the gyrostat considered as a rigid body.

The Hamiltonian (45) expressed in the components of the angular momentum, and after dropping constant terms [44], takes the form

$$\mathcal{H} = \frac{1}{2}(a_1 g_1^2 + a_2 g_2^2 + a_3 g_3^2) - (a_1 f_1 g_1 + a_2 f_2 g_2 + a_3 f_3 g_3), \tag{46}$$

where, as with the rigid body (Section 2), $a_i = 1/I_i$.

Since there are no external forces, the angular momentum vector $\boldsymbol{G}$ is an integral, hence, its norm $\|\boldsymbol{G}\|$ is an integral too:

$$\|\boldsymbol{G}\|^2 = g_1^2 + g_2^2 + g_3^2 = G^2 = \text{constant}; \tag{47}$$

The Poisson structure (6) gives rise to the equations of the motion

$$\begin{aligned}
\dot{g}_1 &= \{g_1\,;\,\mathcal{H}\} = (a_3 - a_2)g_2g_3 + a_2f_2g_3 - a_3f_3g_2,\\
\dot{g}_2 &= \{g_2\,;\,\mathcal{H}\} = (a_1 - a_3)g_1g_3 + a_3f_3g_1 - a_1f_1g_3,\\
\dot{g}_3 &= \{g_3\,;\,\mathcal{H}\} = (a_2 - a_1)g_1g_2 + a_1f_1g_2 - a_2f_2g_1.
\end{aligned} \tag{48}$$

The history of the rotation of $\boldsymbol{G}$ in the body frame is represented as a curve on the S^2 sphere of constant radius G.

The equations (48) admit two integrals, the Hamiltonian (46) and the norm of the total angular momentum (47), therefore the problem is integrable. The phase space of Eqs. (48) may be regarded as a foliation of invariant manifolds

$$S^2(G) = \{(g_1, g_2, g_3)\,|\,g_1^2 + g_2^2 + g_3^2 = G^2\}.$$

Depending on the values of the moments of inertia, that is to say, whether the gyrostat has symmetries or not, and on the number of rotors spinning simultaneously, and also on the axes of inertia about which the rotors spin, the problem is equivalent to the problem of a parametric Hamiltonian on the unit sphere, such as it was proved by the author [44-46]. The number of parameters, and the values of them determines the phase flow.

The differential system (48) belongs to the general class of Hamiltonian systems, the one of the type

$$\mathcal{H} = \mathcal{T}_2 + \mathcal{T}_1, \qquad \text{with} \quad \mathcal{T}_2 = \frac{1}{2}\sum_{1\le i,j\le 3} A_{ij}\xi_i\xi_j \qquad \text{and} \quad \mathcal{T}_1 = \frac{1}{2}\sum_{1\le i\le 3} B_i\xi_i.$$

The unknowns ξ_i $(1 \le i \le 3)$ have the Poisson structure (6).

After the scaling of length and time by

$$(\xi_1, \xi_2, \xi_3) = \frac{1}{G}(g_1, g_2, g_3), \quad (\varphi_1, \varphi_2, \varphi_3) = \frac{1}{G}(f_1, f_2, f_3), \quad \tau = G\,dt$$

the unknowns ξ are on the unit sphere $\mathcal{S}^2$, and the Hamiltonian becomes

$$\mathcal{H} = \frac{1}{2}(a_1\xi_1^2 + a_2\xi_2^2 + a_3\xi_3^2) - (a_1\varphi_1\xi_1 + a_2\varphi_2\xi_2 + a_3\varphi_3\xi_3). \tag{49}$$

As an illustration, let us consider the case of an asymmetric gyrostat $(a_1 > a_2 > a_3)$ with two rotors spinning about any of the two of the principal axes of inertia.

Let us begin with by assuming $\varphi_3 = 0$. By means of the equivalence transformations [47], that is to say, time scaling or adding a constant to the Hamiltonian, it is possible to shift one of the eigenvalues to 0 and to scale one of the two remaining to 1, without changing the symplectic structure of the unknowns. Indeed, the additive transformation

$$\mathcal{H} \longmapsto \mathcal{H} - \frac{a_3}{2}(\xi_1^2 + \xi_2^2 + \xi_3^2),$$

and the scaling transformation

$$\mathcal{H} \longmapsto \mathcal{H}/(a_1 - a_3)$$

give the Hamiltonian

$$\mathcal{H} = \frac{1}{2}\xi_2^2 + \frac{1}{2}\frac{a_2 - a_3}{a_1 - a_3}\xi_2^2 - \frac{a_1\varphi_1}{a_1 - a_3}\xi_1 - \frac{a_2\varphi_2}{a_2 - a_3}\xi_2,$$

that is of form of the tri-parametric case

$$\mathcal{H} = \xi_1^2 + P\xi_2^2 + Q\xi_1 + R\xi_2, \tag{50}$$

with the parameters P, Q and R parameters independent of the unknowns, and $P, Q, R \in \mathbb{R}$. But now, since the hypothesis $a_1 > a_2 > a_3$, the parameter P is such that $0 < P < 1$.

When the rotors are spinning about the axes $\boldsymbol{b}_1$, $\boldsymbol{b}_3$, we have $\varphi_2 = 0$. Now with the additive transformation

$$\mathcal{H} \longmapsto \mathcal{H} - \frac{a_2}{2}(\xi_1^2 + \xi_2^2 + \xi_3^2),$$

and the scaling transformation

$$\mathcal{H} \longmapsto \mathcal{H}/(a_1 - a_2),$$

there results

$$\mathcal{H} = \frac{1}{2}\xi_1^2 + \frac{1}{2}\frac{a_3 - a_2}{a_1 - a_2}\xi_3^2 - \frac{a_1\varphi_1}{a_1 - a_2}\xi_1 - \frac{a_2\varphi_2}{a_1 - a_2}\xi_3,$$

that again belongs to the type (50), but now the parameter is $P < 0$.

In analogous way we have that for $\varphi_1, \varphi_2 \neq 0$ and $\varphi_1 = 0$, the resulting Hamiltonian is

$$\mathcal{H} = \frac{1}{2}\xi_2^2 + \frac{1}{2}\frac{a_3 - a_1}{a_2 - a_1}\xi_3^2 - \frac{a_2\varphi_2}{a_2 - a_1}\xi_2 - \frac{a_3\varphi_3}{a_2 - a_1}\xi_3,$$

and the parameter $\mathrm{P} = (a_3 - a_1)/(a_2 - a_1) > 1$.

Thus with the three possible cases of a triaxial gyrostat with only two rotors spinning about two of the principal axes, we recover the whole triparametric case. The bifurcation partition of the parametric space (P, Q, R) and the possible types of bifurcations have been studied in detail in [48]. Let us mention only that the partition is obtained by the intersection of two surfaces, a conoid and a cone, which normal section is an astroid (see Figure 7). For details about the different bifurcations the reader is addressed to the above quoted work [48].

Acknowledgments

The author is indebted to Professor Cid, who motivated the interest for the rigid bodies, to Dr. Deprit for his geometrical insights of the problem, to Dr. Barkin, who gave us the indications about the obtaining of the Fourier expansions in the Jacobian nome that, later on, were implemented in an symbolic processor by Dr. Vallejo. Many enlighting discussions with Dr. Lanchares clarified the business of the parametric bifurcations.

This paper has been supported by the Spanish Ministry of Education and Science (Projects #PB98-1576 and #ESP99-1074-CO2-01) and by the Centre National d'Études Spatiales at Toulouse.

References

[1] M. Shuster. A survey on attitude representations. *Journal of Astronautical Sciences*, 41:439–517, 1993.

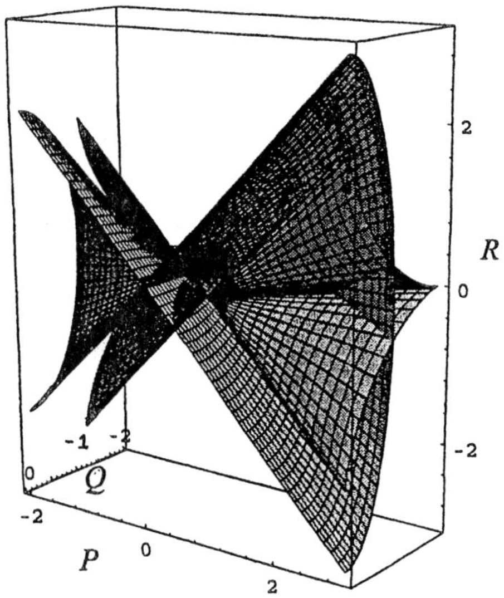

Figure 7: Bifurcation partition of the parametric space (P, Q, R) for the triparametric Hamiltonian

[2] J. Wisdom, J. S. Peale, and F. Mignard. The chaotic rotation of Hyperion. *Icarus*, 58:137–152, 1984.

[3] Iu. A. Sadov. The action-angle variables in the Euler-Poinsot problem. *Akad. Nauka SSSR*, Prep. 22, 1970.

[4] H. Kinoshita. First-order perturbations of the two finite body problem. *Publ. Astron. Soc. Japan*, 24:423–457, 1972.

[5] A. Deprit and A. Elipe. Complete reduction of the Euler-Poinsot problem. *Journal of Astronautical Sciences*, 41:603–628, 1993.

[6] M. Vallejo. *Series de Fourier de funciones elípticas. Aplicación a la precesión terrestre.* PhD thesis, Universidad de Zaragoza, Zaragoza, Spain, 1995.

[7] A. Abad, A. Elipe, and M. Vallejo. Automated Fourier series expansions for elliptic functions. *Mechanics Research Communications*, 21:361–366, 1994.

[8] A. Abad, A. Elipe, and M. Vallejo. Normalization of perturbed non-linear Hamiltonian oscillators. *Int. J. Non-Linear Mechanics*, 32:443–453, 1997.

[9] V. K. Melnikov. On the stability of the center for time periodic perturbations. *Trans. Moscow Math. Soc.*, 12:1–57, 1963.

[10] C. Leubner. Correcting a widespread error concerning the angular velocity of a rotating body. *American Journal of Physics*, 49:232–234, 1981.

[11] P. F. Byrd and M. D. Friedman. *Handbook of elliptic integrals for engineers and scientists.* Springer-Verlag, New York, 1971.

[12] J. A. Serret. Mémoire sur l'emploi de la méthode de la variation des arbitraires dans la théorie des mouvements de rotation. *Mémoires de l' Académie des Sciences de Paris*, 35:585–616, 1866.

[13] H. Andoyer. *Cours de mécanique céleste.* Interdisciplinary Applied Mathematics. Gauthier-Villars et Cie, Paris, 1923.

[14] A. Deprit. Free rotation of a rigid body studied in the phase plane. *American Journal of Physics*, 35:424–428, 1967.

[15] Y. Barkin. *Lecture notes on rigid bodies.* Universidad de Zaragoza. Unpublished, 1992.

[16] I. Y. Bar-Itzhack. New method for extractingthe quaternion from a rotation matrix. *Journal of Guidance, Control and Dynamics*, 23:1085–1087, 2000.

[17] A. Maciejewski. Hamiltonian formalism for Euler parameters. *Celestial Mechanics*, 37:47–57, 1985.

[18] R. Cid and M. E. Sansaturio. Motion of rigid bodies in a set of redundant variables. *Celestial Mechanics and Dynamical Astronomy*, 42:263–274, 1988.

[19] A. Abad, M. Arribas, and A. Elipe. On the attitude of a spacecraft near a Lagrangian point. *Bulletin of the Astronomical Institutes of Czechoslovakia*, 40:302–307, 1989.

[20] H. S. Morton. Hamiltonian and Lagrangian formulation of rigid-body rotational motion based on Euler parameters. *Journal of Astronautical Sciences*, 41:439–517, 1993.

[21] E. Stiefe and G. Scheifele. *Linear and regular Celestial Mechanics.* Springer-Verlag, New York, 1971.

[22] A. Deprit, A. Elipe, and S. Ferrer. Linearization: Laplace versus stiefel. *Celestial Mechanics and Dynamical Astronomy*, 58:151–201, 1994.

[23] E. T. Whittaker. *A treatise on the analytical dynamics of particles and rigid bodies.* Fourth ed. Cambridge Univ. Press, London, 1965.

[24] A. Elipe, A. Abad, and M. Arribas. Scaling Hamiltonians in attitude dynamics of two rigid bodies. *Rev. Acad. Ciencias de Zaragoza*, 47:137–154, 1992.

[25] J. M. Ferrándiz and M. E. Sansaturio. Elimination of the nodes when the satellite is a non-spherical rigid body. *Celestial Mechanics and Dynamical Astronomy*, 46:307–320, 1989.

[26] L. Meirovitch. *Methods of Analytical Dynamics*. McGraw-Hill, New York, 1970.

[27] M. Arribas and A. Elipe. Attitude dynamics of a rigid body on a Keplerian orbit: A simplification. *Celestial Mechanics and Dynamical Astronomy*, 55:243–247, 1993.

[28] A. Deprit. Canonical transformation depending on a small parameter. *Celestial Mechanics*, 1:12–30, 1969.

[29] NEAR. Special issue on the NEAR mission to 433 Eros. *Journal of Astronautical Sciences*, 43:p.1, 1995.

[30] L. Galgani, A. Giorgilli, and J. M. Strelcyn. Chaotic motion and transition tostochasticity in the classical problem of a heavy rigid body with a fixed point. *Il Nuovo Cimento B*, 61:1–20, 1981.

[31] R. Broucke. On the use of Poincaré surfaces of section in rigid-body motion. *Journal of Astronautical Sciences*, 41:593–601, 1993.

[32] V. Lanchares, M. Iñarrea, and J. P. Salas. Spin rotor stabilization of a dual-spin spacecraft with time dependent moments of inertia. *Int. J. Bifurcation and Chaos*, 8:609–617, 1998.

[33] I. S. Gradshtein and I. M. Ryzhik. *Table of integrals, series and products.* Academic Press, San Diego, 1980.

[34] M. Iñarrea and V. Lanchares. Chaos in the reorientation process of a dual-spin spacecraft with time-dependent moments of inertia. *Int. J. of Bifurcation and Chaos*, 10:997–1018, 2000.

[35] V. Volterra. Sur la théorie des variations des latitudes. *Acta Mathematica*, 22:201–358, 1899.

[36] C. H. Hubert. *An attitude adquisition technique for dual-spin spacecraft.* PhD thesis, Cornell University, Ithaca, NY, USA, 1980.

[37] J. E. Cochran, B. H. Shu, and S. D. Rews. Attitude motion of asymmetric ddual-spin spacecraft. *Journal of Guidance, Control and Dynamics*, 5:37–42, 1982.

[38] J. Koiller. A mechanical system with a 'wild' horseshoes. *Journal of Mathematical Physics*, 25:1599–1604, 1984.

[39] C. D. Hall and R. H. Rand. Spinup dynamics of axial dual-spin spacecraft. *Journal of Guidance, Control and Dynamics*, 17:30–37, 1994.

[40] C. D. Hall. Spinup dynamics of biaxial gyrostats. *Journal of Guidance, Control and Dynamics*, 17:30–37, 1994.

[41] V. V. Rumiantsev. On the stability of motion of gyrostats. *Applied Math. and Mech. (PMP)*, 25:9–19, 1961.

[42] A. Elipe, M. Arribas, and A. Riaguas. Complete analysis of bifurcations in the axial gyrostat problem. *Journal of Physics A: Math. and General*, 30:587–601, 1997.

[43] A. Elipe and V. Lanchares. Phase flow of an axially symmetrical gyrostat with one constant rotor. *Journal of Mathematical Physics*, 38:3533–3544, 1997.

[44] A. Elipe and V. Lanchares. Two equivalent problems: Gyrostats in free motion and parametric quadratic Hamiltonian. *Mechanics Research Communications*, 24:583–590, 1997.

[45] A. Elipe. Gyrostats in free rotation. In I. M. Wytrzyszczak, J. H. Lieske, and R. A. Feldman, editors, *Dynamics and Astrometry of Natural and Artificial Celestial Bodies*, pages 391–398. Kluwer, Dordrecht, 1997.

[46] A. Elipe. Hamiltonianos cuadráticos en la esfera unidad o Giróstato en movimiento libre: dos problemas equivalentes. *Rev. Acad. Ciencias de Zaragoza*, 53:5–28, 1998.

[47] J. Frauendiener. Quadratic Hamiltonians on the unit sphere. *Mechanics Research Communications*, 22:313–317, 1995.

[48] V. Lanchares, M. Iñarrea, J. P. Salas, J.D. Sierra, and A. Elipe. Surfaces of bifurcation in a triparametric quadratic Hamiltonian. *Physical Review E*, 52:5540–5548, 1995.

INDEX

S

T

U

V

W

Y

ABOUT THE EDITORS

Thomas F. George was named by the Regents as the 12th Chancellor of the University of Wisconsin-Stevens Point in 1996. Prior to that he served as Provost and Academic Vice President at Washington State University (1991-1996), Dean of Natural Sciences and Mathematics at the State University of New York at Buffalo (1985-1991), and on the chemistry faculty at the University of Rochester (1972-1985) where he was promoted by age 29 to Full professor of Chemistry in 1977. Born in 1947 in Philadelphia, Pennsylvania, Tom George became an Eagle Scout in 1961. He earned his B. A. degree (Phi Beta Kappa) in Mathematics and Chemistry from Gettysburg College in 1967 and his M. S. and Ph.D. degrees from Yale University in 1970, followed by postdoctoral appointments at the Massachusetts Institute of Technology and the University of California at Berkeley. As a faculty member and scientist at the University of Wisconsin-Stevens Point, Dr. George is a Professor of Chemistry and Physics and active researcher in the derivation of theories and development of numerical techniques on computers to describe phenomena in chemical/materials/laser physics. His research has been funded for nearly thirty years from a wide variety of federal agencies and private foundations. He has authored-co-authored 585 articles, 2 graduate-level textbooks, 10 edited books/volumes and 155 conference abstracts. He has delivered numerous invited lectures at international conferences, held lectureships at universities, and presented seminars/colloquia at various institutions and laboratories on science and higher education (315 overall). He maintains active collaborations with scientists around the world and holds the title of Visiting Professor of Physics at Korea University in Seoul. His achievements in the field have garnered awards such as the Marlow Medal and Prize from the Royal Society of Chemistry in Great Britain and Fellowships from the Guggenheim, Sloan and Dreyfus foundations, and he has been named a Fellow of the New York Academy of Sciences, American Physical Society, Society of Photo-Optical Instrumentation Engineers and American Association for the Advancement of Science. He has served on the editorial boards of six different scientific journals and is currently one of the two editors of the *International Journal of Theoretical Physics, Group Theory and Nonlinear Optics*.

Henk F. Arnoldus obtained his Bachelor (1980) and Master (1981) degrees in Physics, both summa cum laude, from Eindhoven University of Technology, The Netherlands, with a specialization in heavy ion scattering theory. He then went to Utrecht University, The Netherlands, where he obtained his Ph.D. degree (1985) in Mathematics and Natural Sciences, with stochastic processes and theoretical quantum optics as research topics. From 1985 to 1988 he was a post-doc with Thomas F. George in the Department of Physics at the University of New York at Buffalo. In 1988 he accepted a faculty position at Villanova University, Pennsylvania, where he stayed for six years. Currently, he is Associate Professor in the Department of Physics and Astronomy at Mississippi State University. His area of research is theoretical optics, with an emphasis on quantum optics, nonlinear optics and near-field optics. He has published 71 papers in peer-reviewed journals, and numerous conference proceedings. He has given 70 seminars/invited talks all over the world, including Russia, Denmark and Korea.